Marx Joyce
Abbott Hardy Machiavelli Chesterton Emerson Austen
Defoe Melville Montaigne Cooper Hugo
Haggard Eliot Grimm
Stoker Carroll Christie Molière
Wilde Maupassant Byron Schiller
Garnett Engels
Goethe Fitzgerald Hawthorne Kafka
Cotton Einstein Dostoyevsky Smith Hall
Baum Kipling Doyle Willis
Leslie Henry Nietzsche
Dumas Flaubert Turgenev Balzac Crane
Stockton Vatsyayana Verne
Burroughs Gogol Vinci
Curtis Tocqueville Whitman Busch
Homer Widger Tolstoy
Darwin Thoreau
Potter Freud Zola Twain Scott Harte
Kant Jowett Lawrence Plato
Andersen Stevenson Dickens Hesse
London Descartes Burton
Poe Aristotle Wells Cervantes
Hale James Hastings Voltaire Cooke
Bunner Shakespeare Irving
Richter Chambers Ida
Doré da Shaw Benedict Alcott
Dante Chekhov Pushkin
Swift Wodehouse Newton

ⓣ tredition®

tredition was established in 2006 by Sandra Latusseck and Soenke Schulz. Based in Hamburg, Germany, tredition offers publishing solutions to authors and publishing houses, combined with world-wide distribution of printed and digital book content. tredition is uniquely positioned to enable authors and publishing houses to create books on their own terms and without conventional manu-facturing risks.

For more information please visit: www.tredition.com

TREDITION CLASSICS

This book is part of the TREDITION CLASSICS series. The creators of this series are united by passion for literature and driven by the intention of making all public domain books available in printed format again - worldwide. Most TREDITION CLASSICS titles have been out of print and off the bookstore shelves for decades. At tredi-tion we believe that a great book never goes out of style and that its value is eternal. Several mostly non-profit literature projects pro-vide content to tredition. To support their good work, tredition donates a portion of the proceeds from each sold copy. As a reader of a TREDITION CLASSICS book, you support our mission to save many of the amazing works of world literature from oblivion. See all available books at www.tredition.com.

ⓖ Project Gutenberg

The content for this book has been graciously provided by Project Gutenberg. Project Gutenberg is a non-profit organization founded by Michael Hart in 1971 at the University of Illinois. The mission of Project Gutenberg is simple: To encourage the creation and distribu-tion of eBooks. Project Gutenberg is the first and largest collection of public domain eBooks.

Handbook of the Trees of New England

Henry M. (Henry Mason) Brooks

Imprint

This book is part of TREDITION CLASSICS

Author: Henry M. (Henry Mason) Brooks
Cover design: Buchgut, Berlin – Germany

Publisher: tredition GmbH, Hamburg - Germany
ISBN: 978-3-8472-3063-2

www.tredition.com
www.tredition.de

HANDBOOK
OF THE
TREES OF NEW ENGLAND

DAME AND BROOKS

CONTENTS

PREFACE.

There is no lack of good manuals of botany in this country. There still seems place for an adequately illustrated book of convenient size for field use. The larger manuals, moreover, cover extensive regions and sometimes fail by reason of their universality to give a definite idea of plants as they grow within more limited areas. New England marks a meeting place of the Canadian and Alleghanian floras. Many southern plants, long after they have abandoned more elevated situations northward, continue to advance up the valleys of the Connecticut and Merrimac rivers, in which they ultimately disappear entirely or else reappear in the valley of the St. Lawrence; while many northern plants pushing southward maintain a more or less precarious existence upon the mountain summits or in the cold swamps of New England, and sometimes follow along the mountain ridges to the middle or southern states. In addition to these two floras, some southwestern and western species have invaded Vermont along the Champlain valley, and thrown out pickets still farther eastward.

At or near the limit of a species, the size and habit of plants undergo great change; in the case of trees, to which this book is restricted, often very noticeable. There is no fixed, absolute dividing line between trees and shrubs. In accordance with the usual definition, a tree must have a single trunk, unbranched at or near the base, and must be at least fifteen feet in height.

Trees that are native in New England, or native in other sections of the United States and thoroughly established in New England, are described and, for the most part, figured. Foreign trees, though locally [Pg iv] established, are not figured. Trees may be occasionally spontaneous over a large area without really forming a constituent part of the flora. Even the apple and pear, when originating spontaneously and growing without cultivation, quickly become degenerate and show little tendency to possess themselves of the soil at the expense of the native growths. Gleditsia, for example, while clearly locally established, has with some hesitation been accorded pictorial representation.

The geographical distribution is treated under three heads: Canada and Alaska; New England; south of New England and westward. With regard to the distribution outside of New England, the standard authorities have been followed. An effort extending through several years has been made to give the distribution as definitely as possible in each of the New England states, and while previous publications have been freely consulted, the present work rests mainly upon the observations of living botanists.

All descriptions are based upon the habit of trees as they appear in New England, unless special mention is made to the contrary. The descriptions are designed to apply to trees as they grow in open land, with full space for the development of their characteristics under favorable conditions. In forest trees there is much greater uniformity; the trunks are more slender, taller, often unbranched to a considerable height, and the heads are much smaller.

When the trunk tapers uniformly from the ground upward, the given diameter is taken at the base; when the trunk is reinforced at the base, the measurements are made above the swell of the roots; when reinforced at the ground and also at the branching point, as often in the American elm, the measurements are made at the smallest place between the swell of the roots and of the branches.

A regular order has been followed in the description for the purpose of ready comparison. No explanation of the headings used seems necessary, except to state that the *habitat* is used in the more customary present acceptation to indicate the place where a plant naturally grows, as in swamps or upon dry hillsides. Under the head of "Horticultural Value," the [Pg v] requisite information is given for an intelligent choice of trees for ornamental purposes.

The order and names of families follow, in the main, Engler and Prantl. In accordance with the general tendency of New England botanists to conform to the best usage until an authoritative agreement has been reached with regard to nomenclature by an international congress, the Berlin rule has been followed for genera, and priority under the genus for species. Other names in use at the present day are given as synonyms and included in the index.

Only those common names are given which are actually used in some part of New England, whether or not the same name is applied to different trees. It seems best to record what is, and not what ought to be. Common names that are the creation of botanists have been disregarded altogether. Any attempt to displace a name in wide use, even by one that is more appropriate, is futile, if not mischievous.

The plates are from original drawings by Mrs. Elizabeth Gleason Bigelow, in all cases from living specimens, and they have been carefully compared with the plates in other works. So far as practicable, the drawings were made of life size, with the exception of the dissected portions of small flowers, which were enlarged. In this way, though not on a perfectly uniform scale, they are, when reduced to the necessary space, distinct in all their parts.

So far as consistent with due precision, popular terms have been used in description, but not when such usage involved tedious periphrase.

Especial mention should be made of those botanists whose assistance has been essential to a knowledge of the distribution of species in the New England states: Maine,—Mr. M. L. Fernald; New Hampshire,—Mr. Wm. F. Flint, Report of Forestry Commission; Vermont,—President Ezra Brainerd; Massachusetts,—trees about Northampton, Mrs. Emily Hitchcock Terry; throughout the Connecticut river valley, Mr. E. L. Morris; Rhode Island,—Professor W. W. Bailey, Professor J. F. Collins; Connecticut,—Mr. C. H. Bissell, Mr. C. K. Averill, Mr. J. N. Bishop. Dr. B. L. [Pg vi] Robinson has given advice in general treatment and in matters of nomenclature; Dr. C. W. Swan and Mr. Charles H. Morss have made a critical examination of the manuscript; Mr. Warren H. Manning has contributed the "Horticultural Values" throughout the work; and Miss M. S. E. James has prepared the index. To these and to all others who have given assistance in the preparation of this work, the grateful thanks of the authors are due. [Pg vii]

[Pg viii]

KEY TO THE TREES OF NEW ENGLAND.
I. LEAVES SIMPLE.

Leaves alternate	A
Outline entire	A C
Outline slightly indented	A D
Outline lobed	A E
Lobes entire	A E F
Lobes slightly indented	A E G
Lobes coarsely toothed	A E H

Leaves opposite	B
A C Ovate to oval, obscurely toothed	Tupelo
A C Ovate to oval	Persimmon
A C Also 3-lobed	Sassafras
A C Sometimes opposite, clustered at the ends of the branchlets	Dogwoods
A D Tremulous habit, oval	Poplars
A D Lanceolate, finely serrate, sometimes entire	Willows
A D Ovate-oval, serrate, doubly serrate	Birches Hornbeams
A D Oval, serrate, oblong-lanceolate, veins terminating in teeth	Beeches Chestnut
A D Ovate-oblong, doubly serrate, surface rough	Elms
A D Ovate to ovate-lanceolate, serrate, surface slightly rough	Hackberry
A D Outline variable, ovate-oval, sometimes lobed (3-7), serrate-dentate	Mulberry

A D Ovate, serrate, oblong	Shadbush Plums Cherries
A D Oval or oval-oblong, spines, evergreen	Holly
A D Broad-ovate, one-sided, serrate	Linden
A D Obovate, oval, lanceolate, oblong	Chestnut oaks
A D Broad-ovate to broad-elliptical, thorny	Thorns
A E F Lobes rounded	Sassafras
A E F Base truncate or heart-shaped	Tulip tree
A E F Obtuse, rounded lobes	White oaks
A E F 3-5-lobed, white-tomentose to glabrous beneath	White poplar
A E G 5-lobed, finely serrate	Sweet gum
A E G Irregularly 3-7-lobed, serrate dentate with equal teeth	Mulberry
A E H Pointed or bristle-tipped lobes	Black oaks
A E H Coarse-toothed or pinnate-lobed, short lobes ending in sharp point	Sycamore
B Outline entire, ovate, veins prominent	Flowering dogwood
B Outline serrate, apex often tapering	Sheep berry
B Outline lobed	Maples

II. LEAVES COMPOUND.

Leaves pinnately compound	I
Leaflets alternate	I A
Outlines of leaflets entire	I A C

	Leaflets opposite	I B
	Leaves bi-pinnately compound	J
I A	Outlines of leaflets with two or three teeth at base.	Ailanthus
I	Outlines of leaflets serrate	Sumacs (except Poison sumac) Mountain ashes Walnuts Hickories
I A C	Leaflets oval, apex obtuse	Locusts (except Honey locust)
I A C	Leaflets oblong, apex acute	Poison sumac
I B	Outlines of leaflets entire	Ashes (except Mountain ashes)
I B	Outlines of leaflets serrate	Ashes (except Mountain ashes)
I B	Leaflets irregularly or coarsely toothed, 3-lobed or nearly entire	Box elder
J	Irregularly bi-pinnate, outlines of leaflets entire, thorns on stem and trunk	Honey locust

[Pg x]

LIST OF PLATES.

[Pg xii]

BOTANICAL AUTHORITIES.

Atkins, C. G.	Pinus Banksiana, Lamb
Averill, C. K.	
	Populus balsamifera, L.
	(*Rhodora*, II, 35)
	Prunus Americana, Marsh.
	Quercus Muhlenbergii, Engelm.
Bailey, L. H.	Populus candicans, Ait.
Bailey, W. W.	Celtis occidentalis, L.
	Fraxinus Pennsylvanica, *var.*
	lanceolata, Sarg.
Bartram, William	Quercus tinctoria (1791)
Batchelder, F. W.	Betula nigra, L.
	Salix discolor, Muhl.
	(Laconia, N. H.)
Bates, J. A.	Pinus Banksiana, Lamb
	Sassafras officinale, Nees
Bishop, J. N.	
	Celtis occidentalis, L.
	Fraxinus Pennsylvanica, Marsh.
	Fraxinus Pennsylvanica, *var.*lanceolata, Sarg.
	Juglans nigra, L. (*in lit.*, 1896)
	Morus rubra, L.
	Populus heterophylla, L.
	Quercus Muhlenbergii, Engelm.

Thuja occidentalis, L.

Bissell, C. H.

Cratægus Crus-Galli, L.

Pinus sylvestris, L. (*in lit.*, 1899)

Prunus Americana, Marsh. (*in lit.*, 1900)

Rhus copallina

Brainerd, Ezra   Carya porcina, Nutt.

Cratægus punctata, Jacq.

Ulmus racemosa, Thomas

Brewster, William   Pinus Banksiana, Lamb

Britton, Nathaniel
Lord   Acer Saccharum, *var.* nigrum

Browne, D. T.   Ilex opaca (*Trees of North America*, 1846)

*Bulletin Torrey Bota-
nical Club,*   XVIII, 150

Pinus Banksiana, Lamb

Chamberlain, E. B.   Ulmus fulva, Michx. (1898)

Churchill, J. R.   Prunus Americana, Marsh.

Collins, J. F.

Gleditsia triacanthos, L.

Dame. L. L.   Cratægus Crus-Galli, L.

Salix fragilis, L. (*Typical Elms and other
Trees of Massachusetts*, p.85

Day, F. M.   Pinus Banksiana, Lamb

Deane, Walter   Sassafras officinale, Nees (1895)

Dudley, W. R.   Populus heterophylla, L.

Eggleston, W.W.   Carya porcina, Nutt.

Celtis occidentalis, L.

Morus rubra, L.

Platanus occidentalis, L.

Populus deltoides, Marsh.

Sassafras officinale, Nees.

Ulmus racemosa, Thomas.

Engler, Adolph

Fernald, M. L.   Fraxinus Pennsylvania, Marsh, *var.* lanceolata, Sarg. (*in lit.*, Sept., 1901)

Gleditsia triacanthos, L.

Populus balsamifera, L. *var.* candicans,

Gray (*Rhodora*. III, 233)

Salix balsamifera, Barratt.

Salix discolor, Muhl. (*in lit.*, Sept., 1901)

Flagg   Morus rubra, L.

Flint, W. F.

Acer Negundo, L.

Quercus alba, L.

Flora of Vermont   Betula lenta, L. (1900)

Cratægus Crus-Galli, L. (1900)

Fraxinus Pennsylvanica, Marsh. (1900)

Picea nigra, Link (1900)

Pinus rigida, Mill (1900)

Populus deltoides, Marsh. (1900)

Quercus alba, L. (1900)

Furbish, Miss Kate   Cratægus coccinea, L. (May, 1899)

	Pinus Banksiana, Lamb
Goodale, G. L.	Pinus Banksiana. Lamb
Grant	Sassafras officinale, Nees
Gray, Asa	Ilex opaca, Ait. (*Manual of Botany*, 6th ed.)
Haines, Mrs.	Pinus Banksiana, Lamb
Harger, E. B.	Picea nigra (*Rhodora*, II, 126)
Harper, R. M.	Liriodendron Tulipifera, L. (*Rhodora* II, 122)
Harrington, A. K.	Picea alba, Link
Haskins, T. H.	Ulmus racemosa, Thomas (*Garden and Forest*, V, 86)
Holmes, Dr. Ezekiel	Nyssa sylvatica, Marsh
Hosford, F. H.	Cratægus mollis, Scheele
Hoyt, Miss Fanny E.	Pinus Banksiana, Lamb
Humphrey, J. E.	Picea alba, Link
	Quercus palustris, Du Roi (*Amherst Trees*)
Jack, J. G.	Cratægus coccinea, L. (1899-1900)
Jessup, Henry Griswold	Carya amara, Nutt
	Ulmus racemosa, Thomas
Josselyn, John	Sassafras officinale, Nees (*New England Rarities*, 1672)
Knowlton, C. H.	Pinus rigida, Mill. (*Rhodora*, II, 124)
Manning, Warren H.	
Matthews, F. Schuyler	Morus rubra. L.

Michaux, fils, François André	Ulmus fulva (*Sylva of North America*, III, ed. 1853)
Morris, E. L.	
Morss, Charles H.	
Oakes, William	Morus rubra, L.
Parlin, J. C.	Sassafras officinale, Nees (1896)
Prantl, Karl von	
Pringle, C. G.	Pinus Banksiana, Lamb
	Pyrus sambucifolia, Cham. & Schlecht
	Quercus Muhlenbergii, Engelm
Rand, E. L.	Pinus Banksiana
Rhodora, III, 234	Acer Saccharum, Marsh., *var.* barbatum, Trelease
	Acer Saccharum, Marsh., *var.* nigrum, Britton
Rhodora, III, 58	Ilex opaca, Ait.
Rhodora, III, 234	Prunus Americana, Marsh
Robbins, James W.	Sassafras officinale, Nees
	Ulmus racemosa, Thomas
Robinson, Dr. B. L.	
Robinson, John	Cratægus coccinea, L. (1900)
Robinson, R. E.	Pinus Banksiana, Lamb
Russell, L. W.	Quercus palustris, Du Roi
	Quercus stellata. Wang
Sargent, Charles S.	Cratægus coccinea, L. (*Botanical Gazette*, XXXI, 12, 1901, by permission)
	Cratægus mollis, Scheele (*Botanical Ga-*

zette. XXXI, 7, 223, 1901)

Setchell, W. A.	Populus heterophylla. L.
Stone, W. E.	Quercus palustris. Du Roi (*Bull. Torr. Club*, IX, 57)
Swan, Dr. C. W.	
Terry, Mrs. Emily H.	Picea alba. Link
Trelease, William	Acer Saccharum, Marsh., *var.* barbatum
Tuckerman, Edward	Betula papyrifera, *var.* minor, Marsh.
Waghorne, A. C.	Cratægus coccinea, L. (1894)

[Pg xvii]

ABBREVIATIONS.

Ait.--Aiton, William.

Barratt, Joseph.

B. S. P.--Britton, Nathaniel Lord, Sterns, E. E., and Poggenburg, Justus F.

Borkh.--Borkhausen, M. B.

Carr.--Carrière, Éli Abel.

Cham.--Chamisso, Adelbert von.

Coulter, John Merle.

DC.--DeCandolle, Augustin Pyramus.

Desf.--Desfontaines, René Louiche.

Du Roi, Johann Philip.

Ehrh.--Ehrhart, Friedrich.

Engelm.--Engelmann, George.

Gray, Asa.

Jacq.--Jacquin, Nicholaus Joseph.

Karst.--Karsten, Hermann Gustav Karl Wilhelm.

Koch, Wilhelm Daniel Joseph.

L.--Linnæus, Carolus.

L. f.--Linnæus, fils, Carl von.

Lam.--Lamarck, J. B. P. A. de Monet.

Lamb, Aylmer Bourke.

Link, Heinrich Friedrich.

Marsh.--Marshall, Humphrey.

Medic.--Medicus, Friedrich Casimir.

Michx.--Michaux, André.

Michaux, fils.--François André.

Mill.--Miller, Philip.

Moench, Konrad.

Muhl.--Muhlenberg, H. Ernst.

Nees--Nees von Esenbeck, C. G.

Nutt.--Nuttall, Thomas.

Peck, Charles H.

Poggenburg, Justus F.

Pursh, Friedrich Trangott.

Roem.--Roemer, Johann Jacob.

Sarg.--Sargent, Charles S.

Scheele, A.

Schlecht--Schlechtendal, D. F. L. von.

Schr.--Schrader, Heinrich A.

Spach, Eduard.

Sterns, E. E.

Sudw.--Sudworth, George B.

Sweet, Robert.

T. and G.--Torrey, John, and Gray, Asa.

Thomas, David.

Vent.--Ventenat, Étienne Pierre.

Walt.--Walter, Thomas.

Wang.--Wangenheim, F. A. J. von.

Watson, Sereno.

Waugh, Frank A.

Willd.--Willdenow, Carl Ludwig.

TREES OF NEW ENGLAND.

PINOIDEÆ. PINE FAMILY. CONIFERS.

ABIETACEÆ. CUPRESSACEÆ.

Trees or shrubs, resinous; leaves simple, mostly evergreen, relatively small, entire, needle-shaped, awl-shaped, linear, or scale-like; stipules none; flowers catkin-like; calyx none; corolla none; ovary represented by a scale (ovuliferous scale) bearing the naked ovules on its surface.

ABIETACEÆ.

Larix. Pinus. Picea. Tsuga. Abies.

Buds scaly; leaves evergreen and persistent for several years (except in *Larix*), scattered along the twigs, spirally arranged or tufted, linear, needle-shaped, or scale-like; sterile and fertile flowers separate upon the same plant; stamens (subtended by scales) spirally arranged upon a central axis, each bearing two pollen-sacs surmounted by a broad-toothed connective; fertile flowers composed of spirally arranged bracts or cover-scales, each bract subtending an ovuliferous scale; cover-scale and ovuliferous scale attached at their bases; cover-scale usually remaining small, ovuliferous scale enlarging, especially after fertilization, gradually becoming woody or leathery and bearing two ovules at its base; cones maturing (except in *Pinus*) the first year; ovuliferous scales in fruit usually known as cone-scales; seeds winged; roots mostly spreading horizontally at a short distance below the surface. [Pg 2]

CUPRESSACEÆ.

Thuja. Cupressus. Juniperus.

Leaf-buds not scaly; leaves evergreen and persistent for several years, opposite, verticillate, or sometimes scattered, scale-like, often needle-shaped in seedlings and sometimes upon the branches of older plants; flowers minute; stamens and pistils in separate blossoms upon the same plant or upon different plants; stamens usually bearing 3-5 pollen-sacs on the underside; scales of fertile aments few, opposite or ternate; fruit small cones, or berries formed by coalescence of the fleshy cone-scales; otherwise as in *Abietaceæ*.

Larix Americana, Michx.

Larix laricina, Koch.

Tamarack. Hacmatack. Larch. Juniper.

Habitat and Range. — Low lands, shaded hillsides, borders of ponds; in New England preferring cold swamps; sometimes far up mountain slopes.

Labrador, Newfoundland, and Nova Scotia, west to the Rocky mountains; from the Rockies through British Columbia, northward along the Yukon and Mackenzie systems, to the limit of tree growth beyond the Arctic circle.

Maine, New Hampshire, and Vermont, — abundant, filling swamps acres in extent, alone or associated with other trees, mostly black spruce; growing depressed and scattered on Katahdin at an altitude of 4000 feet; Massachusetts, — rather common, at least northward; Rhode Island, — not reported; Connecticut, — occasional in the northern half of the state; reported as far south as Danbury (Fairfield county).

South along the mountains to New Jersey and Pennsylvania; west to Minnesota.

[Pg 3]

Habit. — The only New England conifer that drops its leaves in the fall; a tree 30-70 feet high, reduced at great elevations to a height of 1-2 feet, or to a shrub; trunk 1-3 feet in diameter, straight, slender; branches very irregular or in indistinct whorls, for the most part nearly horizontal; often ending in long spire-like shoots; branchlets numerous, head conical, symmetrical while the tree is young, especially when growing in open swamps; when old extremely variable, occasionally with contorted or drooping limbs; foliage pale green, turning to a dull yellow in autumn.

Bark. — Bark of trunk reddish or grayish brown, separating at the surface into small roundish scales in old trees, in young trees smooth; season's shoots gray or light brown in autumn.

Winter Buds and Leaves. — Buds small, globular, reddish.

Leaves simple, scattered along the season's shoots, clustered on the short, thick dwarf branches, about an inch long, pale green, needle-shaped; apex obtuse; sessile.

Inflorescence. — March to April. Flowers lateral, solitary, erect; the sterile from leafless, the fertile from leafy dwarf branches; sterile roundish, sessile; anthers yellow: fertile oblong, short-stalked; bracts crimson or red.

Fruit. — Cones upon dwarf branches, erect or inclining upwards, ovoid to cylindrical, ½-¾ of an inch long, purplish or reddish brown while growing, light brown at maturity, persistent for at least a year; scales thin, obtuse to truncate; edge entire, minutely toothed or erose; seeds small, winged.

Horticultural Value. — Hardy in New England; grows in any good soil, preferring moist locations; the formal outline of the young trees becomes broken, irregular, and picturesque with age, making the mature tree much more attractive than the European species common to cultivation. Rarely for sale in nurseries, but obtainable from collectors. To be successfully transplanted, it must be handled when dormant. Propagated from seed.

Note. — The European species, with which the mature plant is often confused, has somewhat longer leaves and larger cones; a form common in cultivation has long, pendulous branches.

[Pg 4]

33

Plate I. — Larix Americana.

1. Branch with sterile and fertile flowers.
2. Sterile flowers.
3. Different views of stamens.
4. Ovuliferous scale with ovules.
5. Fruiting branch.
6. Open cone.
7. Cone-scale with seeds.
8. Leaf.
9. Cross-section of leaf.

PINUS.

The leaves are of two kinds, primary and secondary; the primary are thin, deciduous scales, in the axils of which the secondary leaf-buds stand; the inner scales of those leaf-buds form a loose, deciduous sheath which encloses the secondary or foliage leaves, which in our species are all minutely serrulate.

Pinus Strobus, L.

White Pine.

Habitat and Range. — In fertile soils; moist woodlands or dry uplands.

Newfoundland and Nova Scotia, through Quebec and Ontario, to Lake Winnipeg.

New England, — common, from the vicinity of the seacoast to altitudes of 2500 feet, forming extensive forests.

South along the mountains to Georgia, ascending to 2500 feet in the Adirondacks and to 4300 in North Carolina; west to Minnesota and Iowa.

Habit. — The tallest tree and the stateliest conifer of the New England forest, ordinarily from 50 to 80 feet high and 2-4 feet in diameter at the ground, but in northern New England, where patches of the primeval forest still remain, attaining a diameter of 3-7 feet and a height ranging from 100 to 150 feet, rising in sombre majesty far above its deciduous neighbors; trunk straight, tapering very gradually; branches nearly horizontal, wide-spreading, in young trees in whorls usually of five, the whorls becoming more or less indistinct in old trees; [Pg 5] branchlets and season's shoots slender; head cone-shaped, broad at the base, clothed with soft, delicate, bluish-green foliage; roots running horizontally near the surface, taking firm hold in rocky situations, extremely durable when exposed.

Bark. — On trunks of old trees thick, shallow-channeled, broad-ridged; on stems of young trees and upon branches smooth, greenish; season's shoots at first rusty-scurfy or puberulent, in late autumn becoming smooth and light russet brown.

Winter Buds and Leaves. — Leading branch-buds ¼-½ inch long, oblong or ovate-oblong, sharp-pointed; scales yellowish-brown.

Foliage leaves in clusters of five, slender, 3-5 inches long, soft bluish-green, needle-shaped, 3-sided, mucronate, each with a single fibrovascular bundle, sessile.

Inflorescence. — June. Sterile flowers at the base of the season's shoots, in clusters, each flower about one inch long, oval, light brown; stamens numerous; connectives scale-like: fertile flowers near the terminal bud of the season's shoots, long-stalked, cylindrical; scales pink-margined.

Fruit. — Cones, 4-6 inches long, short-stalked, narrow-cylindrical, often curved, finally pendent, green, maturing the second year; scales rather loose, scarcely thickened at the apex, not spiny; seeds winged, smooth.

Horticultural Value. — Hardy throughout New England; free from disease; grows well in almost any soil, but prefers a light fertile loam; in open ground retains its lower branches for many years. Good plants, grown from seed, are usually readily obtainable in nurseries; small collected plants from open ground can be moved in sods with little risk.

Several horticultural forms are occasionally cultivated which are distinguished by variations in foliage, trailing branches, dense and rounded heads, and dwarfed or cylindrical habits of growth. [Pg 6]

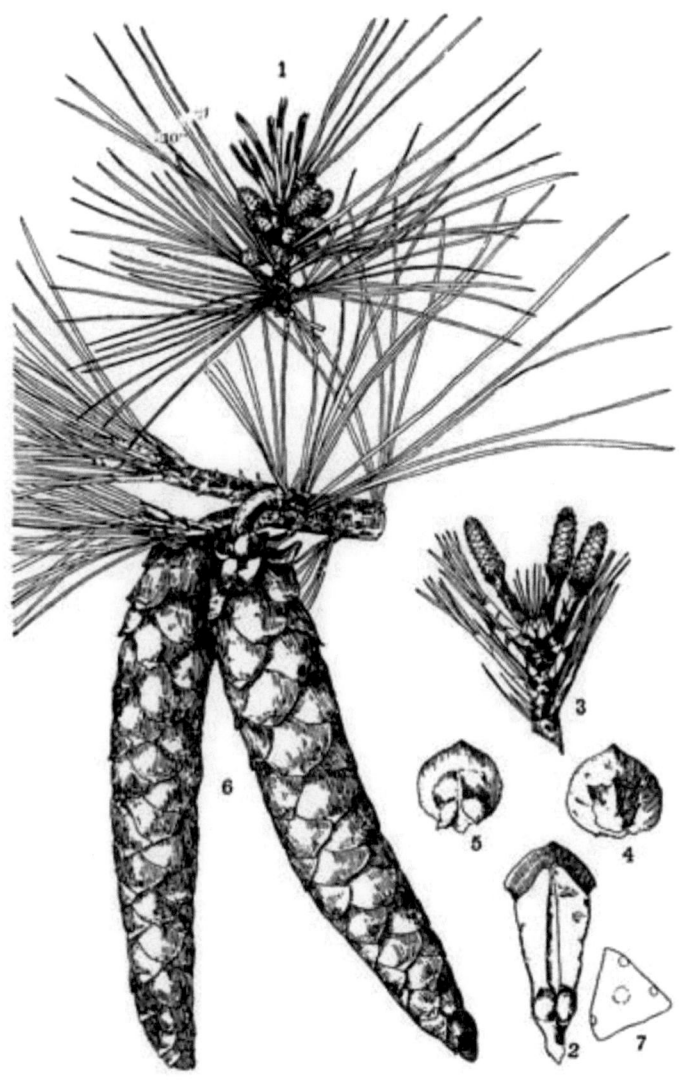

Plate II. Pinus Strobus.

1. Branch with sterile flowers.
2. Stamen.
3. Branch with fertile flowers.
4. Bract and ovuliferous scale, outer side.
5. Ovuliferous scale with ovules, inner side.
6. Branch with cones.
7. Cross-section of leaf.

Pinus rigida, Mill.

Pitch Pine. Hard Pine.

Habitat and Range.—Most common in dry, sterile soils, occasional in swamps.

New Brunswick to Lake Ontario.

Maine,—mostly in the southwestern section near the seacoast; as far north as Chesterville, Franklin county (C. H. Knowlton, *Rhodora*, II, 124); scarcely more than a shrub near its northern limits; New Hampshire,—most common along the Merrimac valley to the White mountains and up the Connecticut valley to the mouth of the Passumpsic, reaching an altitude of 1000 feet above the sea level; Vermont,—common in the northern Champlain valley, less frequent in the Connecticut valley (*Flora of Vermont*, 1900); common in the other New England states, often forming large tracts of woodland, sometimes exclusively occupying extensive areas.

South to Virginia and along the mountains to northern Georgia; west to western New York, Ohio, Kentucky, and Tennessee.

Habit.—Usually a low tree, from 30 to 50 feet high, with a diameter of 1-2 feet at the ground, but not infrequently rising to 70-80 feet, with a diameter of 2-4 feet; trunk straight or more or less tortuous, tapering rather rapidly; branches rising at a wide angle with the stem, often tortuous, and sometimes drooping at the extremities, distinctly whorled in young trees, but gradually losing nearly every trace of regularity; roughest of our pines, the entire framework rough at every stage of growth; head variable, open, often scraggly, widest near the base and sometimes dome-shaped in young trees; branchlets stout, terminating in rigid, spreading tufts of foliage.

[Pg 7]

Bark. — Bark of trunk in old trees thick, deeply furrowed, with broad connecting ridges, separating on the surface into coarse dark grayish or reddish brown scales; younger stems and branches very rough, separating into scales; season's shoots rough to the tips.

Winter Buds and Leaves. — Leading branch-buds ½-¾ inch long, narrow-cylindrical or ovate, acute at the apex, resin-coated; scales brownish.

Foliage leaves in threes, 3-5 inches long, stout, stiff, dark yellowish-green, 3-sided, sharp-pointed, with two fibrovascular bundles; sessile; sheaths when young about ½ inch long.

Inflorescence. — Sterile flowers at the base of the season's shoots, clustered; stamens numerous; anthers yellow: fertile flowers at a slight angle with and along the sides of the season's shoots, single or clustered.

Fruit. — Cones lateral, single or in clusters, nearly or quite sessile, finally at right angles to the stem or twisted slightly downward, ovoid, ovate-conical; subspherical when open, ripening the second season; scales thickened at the apex, armed with stout, straight or recurved prickles.

Horticultural Value. — Hardy throughout New England; well adapted to exposed situations on highlands or along the seacoast; grows in almost any soil, but thrives best in sandy or gravelly moist loams; valuable among other trees for color-effects and occasional picturesqueness of outline; mostly uninteresting and of uncertain habit; subject to the loss of the lower limbs, and not readily transplanted; very seldom offered in quantity by nurserymen; obtainable from collectors, but collected plants are seldom successful. Usually propagated from the seed.

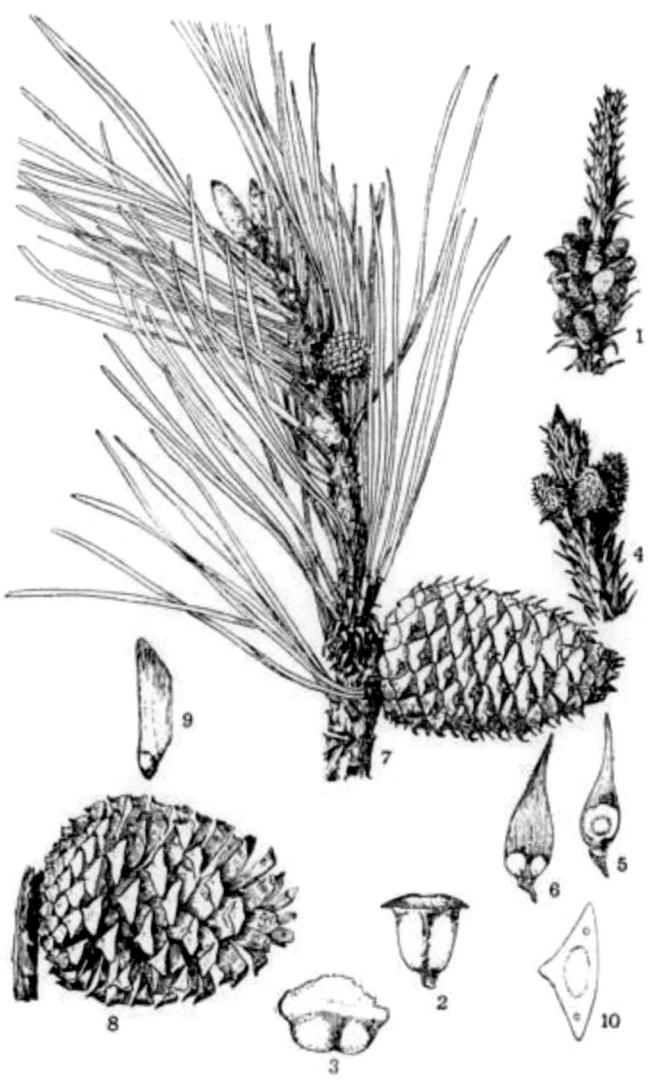

Plate III.—Pinus rigida.

1. Branch with sterile flowers.
2. Stamen, front view.
3. Stamen, top view.
4. Branch with fertile flowers.
5. Fertile flower showing bract and ovuliferous scale, outer side.
6. Fertile flower showing ovuliferous scale with ovules, inner side.
7. Fruiting branch with cones one and two years old.
8. Open cone.
9. Seed.
10. Cross-section of leaf.

[Pg 8]

Pinus Banksiana, Lamb.

Pinus divaricata. Sudw.

Scrub Pine. Gray Pine. Spruce Pine. Jack Pine.

Habitat and Range. — Sterile, sandy soil: lowlands, boggy plains, rocky slopes.

Nova Scotia, northwesterly to the Athabasca river, and northerly down the Mackenzie to the Arctic circle.

Maine, — Traveller mountain and Grand lake (G. L. Goodale); Beal's island on Washington county coast, Harrington, Orland, and Cape Rosier (C. G. Atkins); Schoodic peninsula in Gouldsboro, a forest 30 feet high (F. M. Day, E. L. Rand, *et al.*); Flagstaff (Miss Kate Furbush); east branch of Penobscot (Mrs. Haines); the Forks (Miss Fanny E. Hoyt); Lake Umbagog (Wm. Brewster); New Hampshire, — around the shores of Lake Umbagog, on points extending into the lake, rare (Wm. Brewster *in lit.*, 1899); Welch mountains (*Bull. Torr. Bot. Club*, XVIII, 150); Vermont, — rare, but few trees at each station; Monkton in Addison county (R. E. Robinson); Fairfax, Franklin county (Bates); Starkesboro (Pringle).

West through northern New York, northern Illinois, and Michigan to Minnesota.

Habit. — Usually a low tree, 15-30 feet high and 6-8 inches in diameter at the ground, but under favorable conditions, as upon the wooded points and islands of Lake Umbagog, attaining a height of 50-60 feet, with a diameter of 10-15 inches. Extremely variable in habit. In thin soils and upon bleak sites the trunk is for the most part crooked and twisted, the head scrubby, stunted, and variously distorted, resembling in shape and proportions the pitch pine under similar conditions. In deeper soils, and in situations protected from the winds, the stem is erect, slender, and tapering, surmounted by a stately head with long, flexible branches, scarcely less regular in outline than the spruce. Foliage yellowish-green, bunched at the ends of the branchlets. [Pg 9]

Bark. — Bark of trunk in old trees dark brown, rounded-ridged, rough-scaly at the surface; branchlets dark purplish-brown, rough with the persistent bases of the fallen leaves; season's shoots yellowish-green, turning to reddish-brown.

Winter Buds and Leaves. — Branch-buds light brown, ovate, apex acute or rounded, usually enclosed in resin.

Leaves in twos, divergent from a short close sheath, about 1 inch in length and scarcely 1/12 inch in width, yellowish-green, numerous, stiff, curved or twisted, cross-section showing two fibrovascular bundles; outline narrowly linear; apex sharp-pointed; outer surface convex, inner concave or flat.

Inflorescence. — June. Sterile flowers at the base of the season's shoots, clustered, oblong-rounded: fertile flowers along the sides or about the terminal buds of the season's shoots, single, in twos or in clusters; bracts ovate, roundish, purplish.

Fruit. — Cones often numerous, 1-2 inches long, pointing in the general direction of the twig on which they grow, frequently curved at the tip, whitish-yellow when young, and brown at maturity; scales when mature without prickles, thickened at the apex; outline very irregular but in general oblong-conical. The open cones, which are usually much distorted, with scales at base closed, have a similar outline.

Horticultural Value. — Hardy in New England; slow growing and hard to transplant; useful in poor soil; seldom offered by nurserymen or collectors. Propagated from seed.

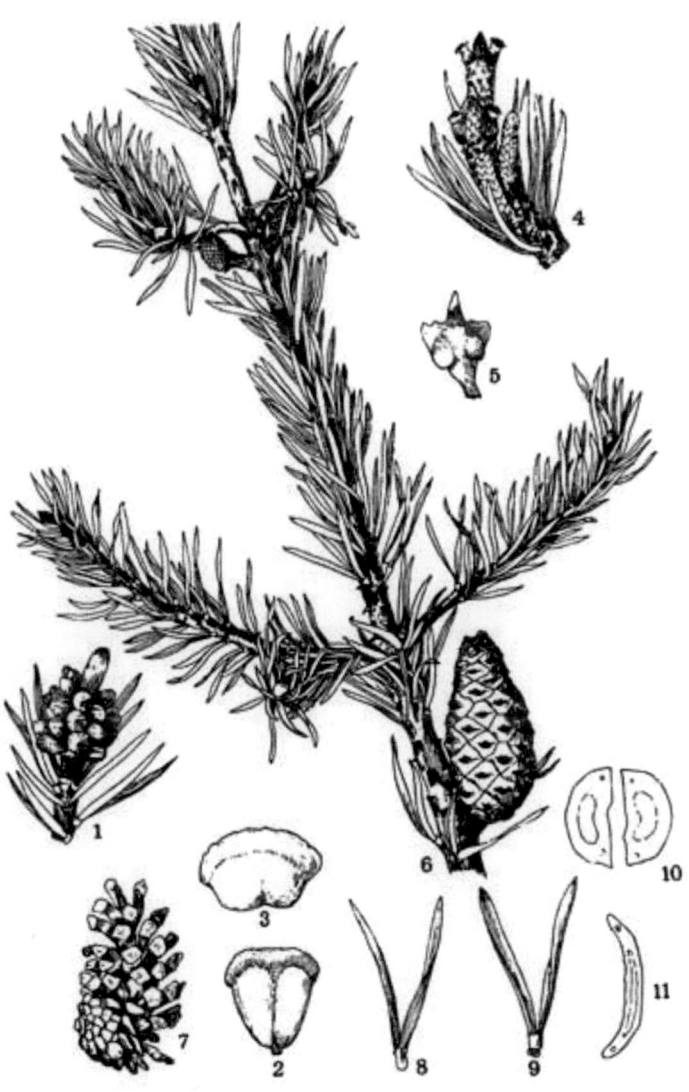

47

Plate IV. — Pinus Banksiana.

1. Branch with sterile flowers.
2. Stamen, front view.
3. Stamen, top view.
4. Branch with fertile flowers.
5. Ovuliferous scale with ovules, inner side.
6. Fruiting branch.
7. Open cone.
8, 9. Variant leaves.
10, 11. Cross-sections of leaves.

[Pg 10]

Pinus resinosa, Ait.

Red Pine. Norway Pine.

Habitat and Range. — In poor soils: sandy plains, dry woods.

Newfoundland and New Brunswick, throughout Quebec and Ontario, to the southern end of Lake Winnipeg.

Maine, — common, plains, Brunswick (Cumberland county); woods, Bristol (Lincoln county); from Amherst (western part of Hancock county) and Clifton (southeastern part of Penobscot county) northward just east of the Penobscot river the predominant tree, generally on dry ridges and eskers, but in Greenbush and Passadumkeag growing abundantly on peat bogs with black spruce; hillsides and lower mountains about Moosehead, scattered; New Hampshire, — ranges with the pitch pine as far north as the White mountains, but is less common, usually in groves of a few to several hundred acres in extent; Vermont, — less common than *P. Strobus* or *P. rigida*, but not rare; Massachusetts, — still more local, in stations widely separated, single trees or small groups; Rhode Island, — occasional; Connecticut, — not reported.

South to Pennsylvania; west through Michigan and Wisconsin to Minnesota.

Habit. — The most beautiful of the New England pines, 50-75 feet high, with a diameter of 2-3 feet at the ground; reaching in Maine a height of 100 feet and upwards; trunk straight, scarcely tapering; branches low, stout, horizontal or scarcely declined, forming a broad-based, rounded or conical head of great beauty when young, becoming more or less irregular with age; foliage of a rich dark green, in long dense tufts at the ends of the branches.

Bark. — Bark of trunk reddish-brown, in old trees marked by flat ridges which separate on the surface into thin, flat, loose scales; branchlets rough with persistent bases of leaf buds; season's shoots stout, orange-brown, smooth.

Winter Buds and Leaves. — Leading branch-buds conical, [Pg 11] about 3/4 inch long, tapering to a sharp point, reddish-brown, invested with rather loose scales.

Foliage leaves in twos, from close, elongated, persistent, and conspicuous sheaths, about 6 inches long, dark green, needle-shaped, straight, sharply and stiffly pointed, the outer surface round and the inner flattish, both surfaces marked by lines of minute pale dots.

Inflorescence. — Sterile flowers clustered at the base of the season's shoots, oblong, ½-¾ inch long: fertile flowers single or few, at the ends of the season's shoots.

Fruit. — Cones near extremity of shoot, at right angles to the stem, maturing the second year, 1-3 inches long, ovate to oblong conical; when opened broadly oval or roundish; scales not hooked or pointed, thickened at the apex.

Horticultural Value. — Hardy in New England; a tall, dark-foliaged evergreen, for which there is no substitute; grows rapidly in all well-drained soils and in exposed inland or seashore situations; seldom disfigured by insects or disease; difficult to transplant and not common in nurseries. Propagated from seed.

Plate V. — Pinus resinosa.

1. Branch with sterile flowers.
2. Stamen, front view.
3. Stamen, top view.
4. Branch with fertile flowers and one-year-old cones.
5. Bract and ovuliferous scale, outer side.
6. Ovuliferous scale with ovules, inner side.
7. Fruiting branch showing cones of three different seasons.
8. Seeds with cone-scale.
9, 10. Cross-sections of leaves.

Pinus sylvestris, L.

Scotch Pine (sometimes incorrectly called the Scotch fir).

Indigenous in the northern parts of Scotland and in the Alps, and from Sweden and Norway, where it forms large forests eastward throughout northern Europe and Asia.

At Southington, Conn., many of these trees, probably originating from an introduced pine in the vicinity, were [Pg 12] formerly scattered over a rocky pasture and in the adjoining woods, a tract of about two acres in extent. Most of these were cut down in 1898, but the survivors, if left to themselves, will doubtless multiply rapidly, as the conditions have proved very favorable (C. H. Bissell *in lit.*, 1899).

Like *P. resinosa* and *P. Banksiana*, it has its foliage leaves in twos, with neither of which, however, is it likely to be confounded; aside from the habit, which is quite different, it may be distinguished from the former by the shortness of its leaves, which are less than 2 inches long, while those of *P. resinosa* are 5 or 6; and from the latter by the position of its cones, which point outward and downward at maturity, while those of *P. Banksiana* follow the direction of the twig.

Picea nigra, Link.

Picea Mariana, B. S. P. (including Picea brevifolia, Peck).

Black Spruce. Swamp Spruce. Double Spruce. Water Spruce.

Habitat and Range.—Swamps, sphagnum bogs, shores of rivers and ponds, wet, rocky hillsides; not uncommon, especially northward, on dry uplands and mountain slopes.

Labrador, Newfoundland, and Nova Scotia, westward beyond the Rocky mountains, extending northward along the tributaries of the Yukon in Alaska.

Maine,—common throughout, covering extensive areas almost to the exclusion of other trees in the central and northern sections, occasional on the top of Katahdin (5215 feet); New Hampshire and Vermont,—common in sphagnum swamps of low and high altitudes; the dwarf form, var. *semi-prostrata*, occurs on the summit of Mt. Mansfield (*Flora of Vermont*, 1900); Massachusetts,—frequent; Rhode Island,—not reported; Connecticut,—rare; on north shore of Spectacle ponds in Kent (Litchfield county), at an elevation of 1200 feet; Newton (Fairfield county), a few scattered trees in a swamp at an altitude of 400 feet: (New Haven county) a few [Pg 13] small trees at Bethany; at Middlebury abundant in a swamp of five acres (E. B. Harger, *Rhodora*, II, 126).

South along the mountains to North Carolina and Tennessee; west through the northern tier of states to Minnesota.

Habit.—In New England, usually a small, slender tree, 10-30 feet high and 5-8 inches in diameter; attaining northward and westward much greater dimensions; reduced at high elevation to a shrub or dwarf tree, 2 or 3 feet high; trunk tapering very slowly, forming a narrow-based, conical, more or less irregular head; branches rather short, scarcely whorled, horizontal or more frequently declining with an upward tendency at the ends, often growing in open swamps almost to the ground, the lowest prostrate, sometimes rooting at their tips and sending up shoots; spray stiff and rather slender; foliage dark bluish-green or glaucous. This tree often begins to blossom after attaining a height of 2-5 feet, the terminal cones each season remaining persistent at the base of the branches, sometimes for many years.

Bark. — Bark of trunk grayish-brown, separating into rather close, thin scales; branchlets roughened with the footstalks of the fallen leaves; twigs in autumn dull reddish-brown with a minute, erect, pale, rusty pubescence, or nearly smooth.

Winter Buds and Leaves. — Buds scaly, ovate, pointed, reddish-brown. Leaves scattered, needle-shaped, dark bluish-green, the upper sides becoming yellowish in the sunlight, the faces marked by parallel rows of minute bluish dots which sometimes give a glaucous effect to the lower surface or even the whole leaf on the new shoots, 4-angled, ¼-¾ of an inch long, straight or slightly incurved, blunt at the apex, abruptly tipped or mucronate, sessile on persistent, decurrent footstalks.

Inflorescence. — April to May, a week or two earlier than the red spruce; sterile flowers terminal or axillary, on wood of the preceding year; about 3/8 inch long, ovate; anthers madder-red: fertile flowers at or near end of season's shoots, erect; scales madder-red, spirally imbricated, broader than long, margin erose, rarely entire.

Fruit. — Cones, single or clustered at or near ends of the [Pg 14] season's shoots, attached to the upper side of the twig, but turning downward by the twisting of the stout stalk, often persistent for years; ½-1½ inches long; purplish or grayish brown at the end of the first season, finally becoming dull reddish or grayish brown, ovate, ovate-oval, or nearly globular when open; scales rigid, thin, reddish on the inner surface; margin rounded, uneven, eroded, bifid, or rarely entire.

Horticultural Value. — Best adapted to cool, moist soils; of little value under cultivation; young plants seldom preserving the broad-based, cone-like, symmetrical heads common in the spruce swamps, the lower branches dying out and the whole tree becoming scraggly and unsightly. Seldom offered by nurserymen.

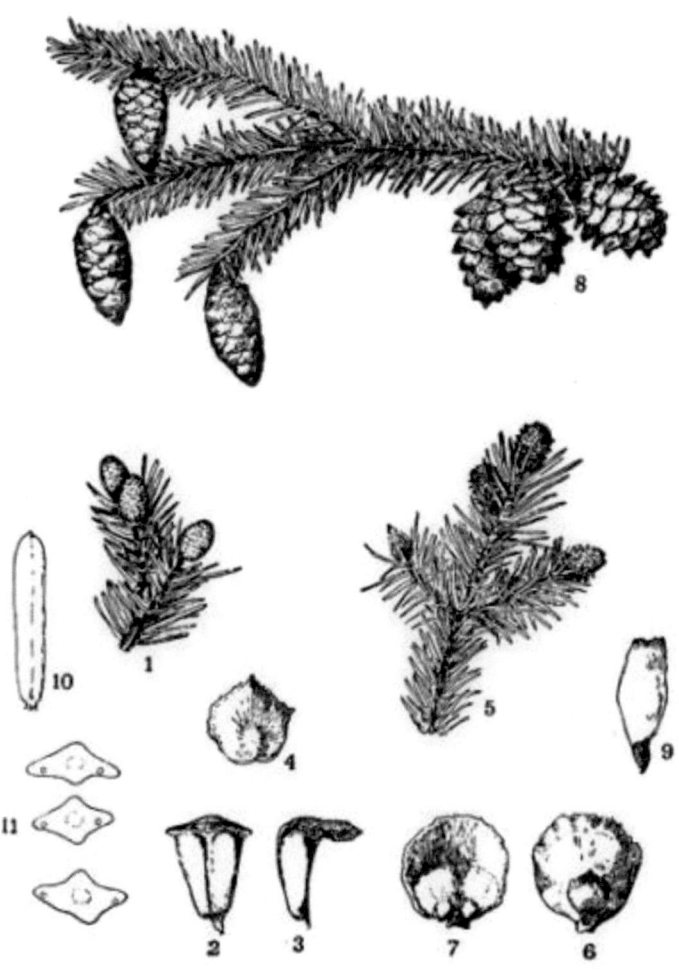

Plate VI. — Picea nigra.

1. Branch with sterile flowers.
2. Stamen, front view.
3. Stamen, side view.
4. Stamen, top view.
5. Branch with fertile flowers.
6. Cover-scale and ovuliferous scale, outer side.
7. Ovuliferous scale with ovules, inner side.
8. Fruiting branch.
9. Seed.
10. Leaf.
11. Cross-sections of leaves.

Picea rubra, Link.

Picea rubens, Sarg. Picea nigra, var. rubra, Engelm.

Red Spruce.

Habitat and Range.—Cool, rich woods, well-drained valleys, slopes of mountains, not infrequently extending down to the borders of swamps.

Prince Edward island and Nova Scotia, along the valley of the St. Lawrence.

Maine,—throughout: most common towards the coast and in the extreme north, thus forming a belt around the central area, where it is often quite wanting except on cool or elevated [Pg 15] slopes; New Hampshire,—throughout; the most abundant conifer of upper Coos, the White mountain region where it climbs to the alpine area, and the higher parts of the Connecticut-Merrimac watershed; Vermont,—throughout; the common spruce of the Green mountains, often in dense groves on rocky slopes with thin soil; Massachusetts,—common in the mountainous regions of Berkshire county and on uplands in the northern sections, occasional southward; Rhode Island and Connecticut,—not reported.

South along the Alleghanies to Georgia, ascending to an altitude of 4500 feet in the Adirondacks, and 4000-5000 feet in West Virginia; west through the northern tier of states to Minnesota.

Habit.—A hardy tree, 40-75 feet high; trunk 1-2½ feet in diameter, straight, tapering very slowly; branches longer than those of the black spruce, irregularly whorled or scattered, the lower often declined, sometimes resting on the ground, the upper rising toward the light, forming while the tree is young a rather regular, narrow, conical head, which in old age and in bleak mountain regions becomes, by the loss of branches, less symmetrical but more picturesque; foliage dark yellowish-green.

Bark.—Bark of trunk smoothish and mottled on young trees, at length separating into small, thin, flat, reddish scales; in old trees striate with shallow sinuses, separating into ashen-white plates, often partially detached; spray reddish or yellowish white in autumn with minute, erect, pale rusty pubescence.

Winter Buds and Leaves. — Buds scaly, conical, brownish, ⅓ inch long. Leaves solitary, at first closely appressed around the young shoots, ultimately pointing outward, those on the underside often twisting upward, giving a brush-like appearance to the twig, ½-¾ inch long, straight or curved (curvature more marked than in *P. nigra*), needle-shaped, dark yellowish-green, 4-angled; apex blunt or more or less pointed, often mucronate; base blunt; sessile on persistent leaf-cushions.

Inflorescence. — May. Sterile flowers terminal or axillary on wood of the preceding year, ½-¾ inch long, cylindrical; anthers pinkish-red: fertile flowers lateral along previous [Pg 16] season's shoots, erect; scales madder-purple, spirally imbricated, broader than long, margin entire or slightly erose.

Fruit. — Cones; single or clustered, lateral along the previous season's shoots, recurved, mostly pointing downward at various angles, on short stalks, falling the first autumn but sometimes persistent a year longer, 1-2 inches long (usually larger than those of *P. nigra*), reddish-brown, mostly ovate; scales thin, stiff, rounded; margin entire or slightly irregular.

Horticultural Value. — Hardy throughout New England; adapts itself to a great variety of soils and lives to a great age. Its narrow-based conical form, dense foliage, and yellow green coloring form an effective contrast with most other evergreens. It grows, however, slowly, is subject to the loss of its lower branches and to disfigurement by insects. Seldom offered in nurseries.

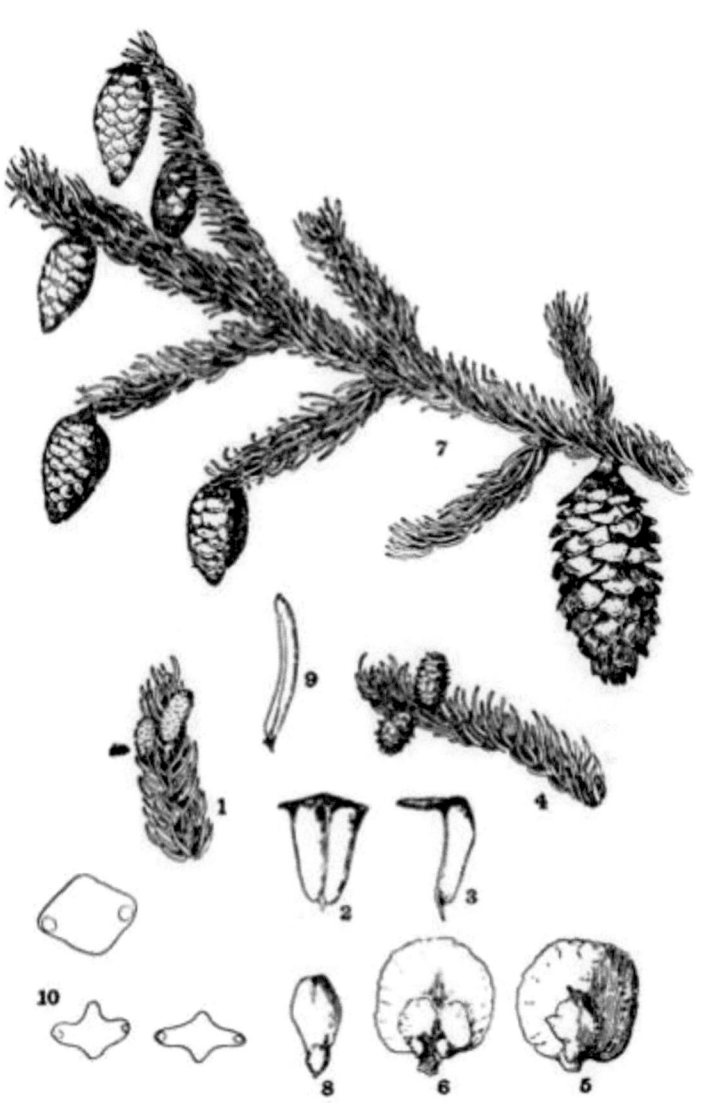

Plate VII.--Picea rubra.

1. Branch with sterile flowers.
2. Stamen, front view.
3. Stamen, side view.
4. Branch with fertile flowers.
5. Cover-scale and ovuliferous scale, outer side.
6. Ovuliferous scale with ovules, inner side.
7. Fruiting branch with cones of two seasons.
9. Seed.
10. Leaf.
11. Cross-sections of leaves.

Picea alba, Link.

Picea Canadensis, B. S. P.

White Spruce. Cat Spruce. Skunk Spruce. [1] Labrador Spruce.

Habitat and Range.--Low, damp, but not wet woods; dry, sandy soils, high rocky slopes and exposed hilltops, often in scanty soil.

> Newfoundland and Nova Scotia, through the provinces of Quebec and Ontario to Manitoba and British Columbia, northward beyond all other trees, within 20 miles of the Arctic sea.

[Pg 17]

Maine,--frequent in sandy soils, often more common than _P. rubra_, as far south as the shores of Casco bay; New Hampshire,-- abundant around the shores of the Connecticut river, disappearing southward at Fifteen-Mile falls; Vermont,--restricted mainly to the northern sections, more common in the northeast; Massachusetts,-- occasional in the mountainous regions of Berkshire county; a few trees in Hancock (A. K. Harrington); as far south as Amherst (J. E. Humphrey) and Northampton (Mrs. Emily H. Terry), probably about the southern limit of the species; Rhode Island and Connecticut,--not reported.

> West through the northern sections of the northern tier of states to the Rocky mountains.

Habit.--A handsome tree, 40-75 feet high, with a diameter of 1-2 feet at the ground, the trunk tapering slowly, throwing out numerous scattered or irregularly whorled, gently ascending or nearly horizontal branches, forming a symmetrical, rather broad conical head, with numerous branchlets and bluish-green glaucous foliage spread in dense planes; gum bitter.

Bark.--Bark of trunk pale reddish-brown or light gray, on very old trees ash-white; not as flaky as the bark of the red spruce, the scales smaller and more closely appressed; young trees and small branches much smoother, pale reddish-brown or mottled brown and gray, resembling the fir balsam; branchlets glabrous; shoots from which the leaves have fallen marked by the scaly, persistent leaf-cushions; new shoots pale fawn-color at first, turning darker the second season; bark of the tree throughout decidedly lighter than that of the red or black spruces.

Winter Buds and Leaves.--Buds scaly, ovoid or conical, about ¼ inch long, light brown. Leaves scattered, stout as those of _P. rubra_ or very slender, those on the lower side straight or twisted so as to appear on the upper side, giving a brush-like appearance to the twig, about ¾ of an inch long; bluish-green, glaucous on the new shoots, needle-shaped, 4-angled, slightly curved, bluntish or sharp-pointed, often mucronate, marked on each side with several parallel rows of dots, malodorous, especially when bruised.

[Pg 19]

Inflorescence.--April to May. Sterile flowers terminal or axillary, on wood of the preceding season; distinctly stalked; cylindrical, 1/2 an inch long; anthers pale red: fertile flowers at or near ends of season's shoots; scales pale red or green, spirally imbricated, broader than long; margin roundish, entire or nearly so; each scale bearing two ovules.

Fruit.--Cones short-stalked, at or near ends of branchlets, light green while growing, pale brownish when mature, spreading, 1-2-1/2 inches long, when closed cylindrical, tapering towards the apex, cylindrical or ovate-cylindrical when open, mostly falling the first winter; scales broad, thin, smooth; margin rounded, sometimes straight-topped, usually entire.

Horticultural Value.--A beautiful tree, requiring cold winters for its finest development, the best of our New England spruces for ornamental and forest plantations in the northern sections; grows rapidly in moist or well-drained soils, in open sun or shade, and in exposed situations. The foliage is sometimes infested by the red spider. Propagated from seed.

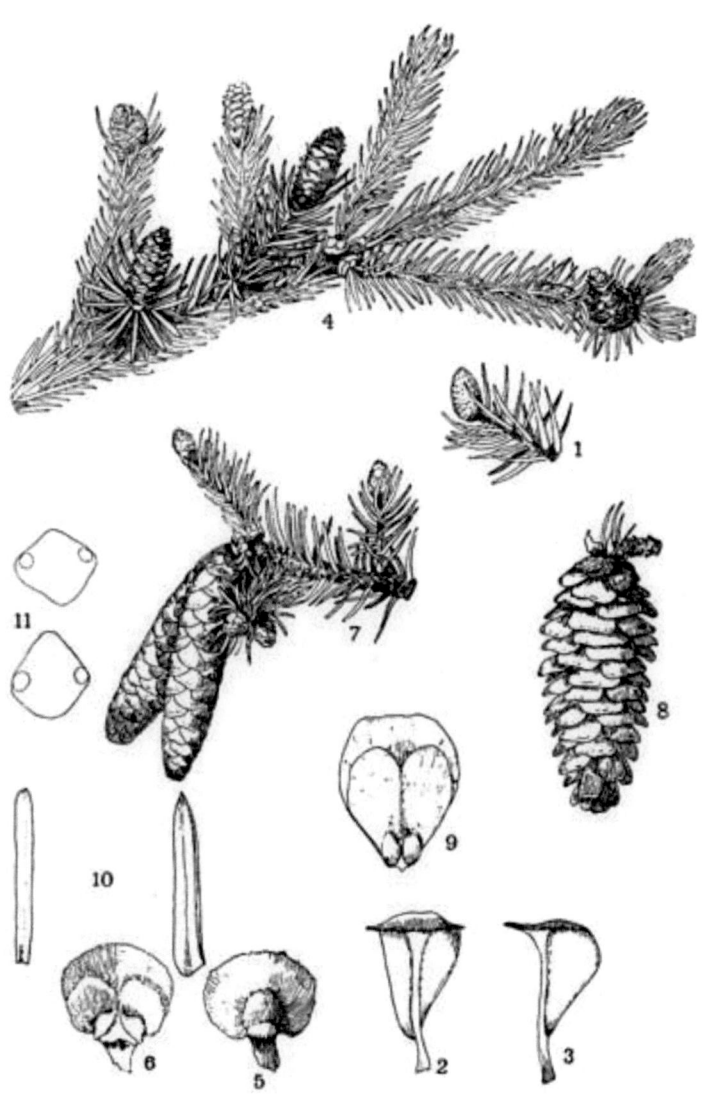

69

Plate VIII.--Picea alba.

[Pg 18]

1. Branch with sterile flowers.
2. Stamen, front view.
3. Stamen, side view.
4. Branch with fertile flowers.
5. Cover-scale and ovuliferous scale, outer side.
6. Ovuliferous scale with ovules, inner side.
7. Fruiting branch.
8. Open cone.
9. Seed with ovuliferous scale.
10. Leaves.
11. Cross-sections of leaves.

[1] So called from the peculiarly unpleasant odor of the crushed foliage and young shoots,--a characteristic which readily distinguishes it from the _P. nigra_ and _P. rubra_.

Tsuga Canadensis, Carr.

Hemlock.

Habitat and Range.—Cold soils, borders of swamps, deep woods, ravines, mountain slopes.

Nova Scotia, New Brunswick, through Quebec and Ontario.

Maine,—abundant, generally distributed in the southern and central portions, becoming rare northward, disappearing entirely in most of Aroostook county and the northern Penobscot region; New Hampshire,—abundant, from the sea to a height of 2000 feet in the White mountains, disappearing in upper Coos county; Vermont,—common, especially in the mountain forests; Massachusetts, Rhode Island, and Connecticut,—common.

South to Delaware and along the mountains to Georgia and Alabama, ascending to an altitude of 2000 feet in the Adirondacks; west to Michigan and Minnesota.

Habit.—A large handsome tree, 50-80 feet high; trunk 2-4 feet in diameter, straight, tapering very slowly; branches going out at right angles, not disposed in whorls, slender, brittle yet elastic, the lowest declined or drooping; head spreading, somewhat irregular, widest at the base; spray airy, graceful, plume-like, set in horizontal planes; foliage dense, extremely delicate, dark lustrous green above and silver green below, tipped in spring with light yellow green.

Bark.—Bark of trunk reddish-brown, interior often cinnamon red, shallow-furrowed in old trees; young trunks and branches of large trees gray brown, smooth; season's shoots very slender, buff or light reddish-brown, minutely pubescent.

Winter Buds and Leaves.—Winter buds minute, red brown. Leaves spirally arranged but brought by the twisting of the leafstalk into two horizontal rows on opposite sides of the twig, about ½ an inch long, yellow green when young, becoming at maturity dark shining green on the upper surface, white-banded along the midrib beneath, flat, linear, smooth, occasionally minutely toothed, especially in the upper half; apex [Pg 20] obtuse, base obtuse; leafstalk slender, short but distinct, resting on a slightly projecting leaf-cushion.

Inflorescence.—Sterile flowers from the axils of the preceding year's leaves, consisting of globose clusters of stamens with spurred anthers: fertile catkins at ends of preceding year's branchlets, scales crimson.

Fruit.—Cones, on stout footstalks at ends of branchlets, pointing downward, ripening the first year, light brown, about 3/4 of an inch long, ovate-elliptical, pointed; scales rounded at the edge, entire or obscurely toothed.

Horticultural Value.—Hardy throughout New England; grows almost anywhere, but prefers a good, light, loamy or gravelly soil on moist slopes; a very effective tree single or in groups, useful in shady places, and a favorite hedge plant; not affected by rust or insect enemies; in open ground retains its lower branches for many years. About twenty horticultural forms, with variations in foliage, of columnar, densely globular, or weeping habit, are offered for sale in nurseries.

Plate IX. — Tsuga Canadensis.

1. Branch with flower-buds.
2. Branch with sterile flowers.
3. Sterile flowers.
4. Spurred anther.
5. Branch with fertile flowers.
6. Ovuliferous scale with ovule, inner side.
7. Fruiting branch.
8. Cover-scales with seeds.
9. Leaf.
10. Cross-section of leaf.

Abies balsamea, Mill.

Fir Balsam. Balsam. Fir.

Habitat and Range. — Rich, damp, cool woods, deep swamps, mountain slopes.

Labrador, Newfoundland, and Nova Scotia, northwest to the Great Bear Lake region.

Maine, — very generally distributed, ordinarily associated with white pine, black spruce, red spruce, and a few deciduous trees, growing at an altitude of 4500 feet upon Katahdin; New Hampshire, — common in upper Coos county and in the White mountains, where it climbs up to the alpine area; in [Pg 21] the southern part of the state, in the extensive swamps around the sources of the Contoocook and Miller's rivers, it is the prevailing timber; Vermont, — common; not rare on mountain slopes and even summits; Massachusetts, — not uncommon on mountain slopes in the northwestern and central portions of the state, ranging above the red spruces upon Graylock; a few trees here and there in damp woods or cold swamps in the southern and eastern sections, where it has probably been accidentally introduced; Rhode Island and Connecticut, — not reported.

South to Pennsylvania and along high mountains to Virginia; west to Minnesota.

Habit. — A slender, handsome tree, the most symmetrical of the New England spruces, with a height of 25-60 feet, and a diameter of 1-2 feet at the ground, reduced to a shrub at high altitudes; branches in young trees usually in whorls; branchlets mostly opposite. The branches go out from the trunk at an angle varying to a marked degree even in trees of about the same size and apparent age; in some trees declined near the base, horizontal midway, ascending near the top; in others horizontal or ascending throughout; in others declining throughout like those of the Norway spruce; all these forms growing apparently under precisely the same conditions; head widest at the base and tapering regularly upward; foliage dark bright green; cones erect and conspicuous.

Bark. — Bark of trunk in old trees a variegated ashen gray, appearing smooth at a short distance, but often beset with fine scales, with one edge scarcely revolute, giving a ripply aspect; branches and young trees mottled or striate, greenish-brown and very smooth; branchlets from which the leaves have fallen marked with nearly circular leaf-scars; season's shoots pubescent; bark of trunk in all trees except the oldest with numerous blisters, containing the Canada balsam of commerce.

Winter Buds and Leaves. — Buds small, roundish, resinous, grouped on the leading shoots. Leaves scattered, spirally arranged in rows, at right angles to twig, or disposed in two [Pg 22] ranks like the hemlock;*nbsp;½-1 inch long, dark glossy green on the upper surface, beneath silvery bluish-white, and traversed lengthwise by rows of minute dots, flat, narrowly linear; apex blunt, in young trees and upon vigorous shoots, often slightly but distinctly notched, or sometimes upon upper branches with a sharp, rigid point; sessile; aromatic.

Inflorescence. — Early spring. Lateral or terminal on shoots of the preceding season; sterile flowers oblong-cylindrical, ¼ inch in length; anthers yellow, red-tinged: fertile flowers on the upper side of the twig, erect, cylindrical; cover-scales broad, much larger than the purple ovuliferous scales, terminating in a long, recurved tip.

Fruit.—Cones along the upper side of the branchlets, erect or nearly so in all stages of growth, purplish when young, 3-5 inches long, 1 inch or more wide; puberulous; cover-scales at maturity much smaller than ovuliferous scales, thin, obovate, serrulate, bristle-pointed; ovuliferous scales thin, broad, rounded; edge minutely erose, serrulate or entire; both kinds of scales falling from the axis at maturity; seeds winged, purplish.

Horticultural Value.—Hardy in New England, but best adapted to the northern sections; grows rapidly in open or shaded situations, especially where there is cool, moist, rich soil; easily transplanted; suitable for immediate effects in forest plantations, but not desirable for a permanent ornamental tree, as it loses the lower branches at an early period. Nurserymen and collectors offer it in quantity at a low price. Propagated from seed.

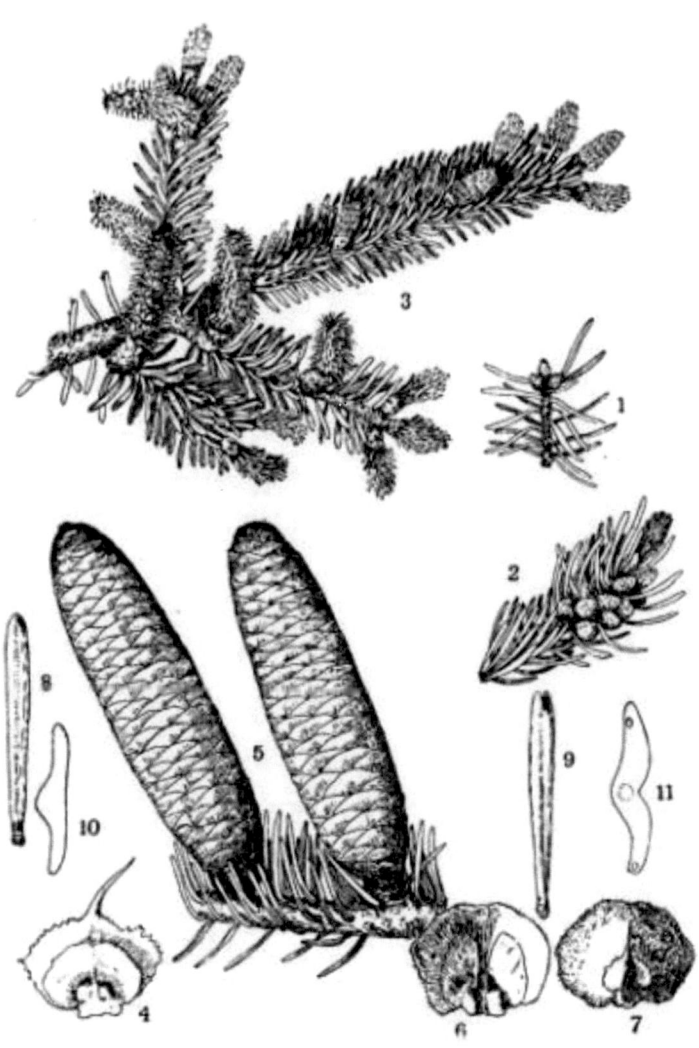

Plate X. — Abies balsamea.

1. Branch with flower-buds.
2. Branch with sterile flowers.
3. Branch with fertile flowers.
4. Cover-scale and ovuliferous scale with ovules, inner side.
5. Fruiting branch.
6. Ovuliferous scales with ovules at maturity, inner side.
7. Cone-scale and ovuliferous scale at maturity, outer side.
8-9. Leaves.
10-11. Cross-sections of leaves.

[Pg 23]

Thuja occidentalis, L.

Arbor-Vitæ. White Cedar. Cedar.

Habitat and Range.—Low, swampy lands, rocky borders of rivers and ponds.

Southern Labrador to Nova Scotia; west to Manitoba.

Maine,—throughout the state; most abundant in the central and northern portions, forming extensive areas known as "cedar swamps"; sometimes bordering a growth of black spruce at a lower level; New Hampshire,—mostly confined to the upper part of Coos county, disappearing at the White river narrows near Hanover; seen only in isolated localities south of the White mountains; Vermont,— common in swamps at levels below 1000 feet; Massachusetts,— Berkshire county; occasional in the northern sections of the Connecticut river valley; Rhode Island,—not reported; Connecticut,— East Hartford (J. N. Bishop).

South along the mountains to North Carolina and East Tennessee; west to Minnesota.

Habit.—Ordinarily 25-50 feet high, with a trunk diameter of 1-2 feet, in northern Maine occasionally 60-70 feet in height, with a diameter of 3-5 feet; trunk stout, more or less buttressed in old trees, tapering rapidly, often divided, inclined or twisted, ramifying for the most part near the ground, forming a dense head, rather small for the size of the trunk; branches irregularly disposed and nearly horizontal, the lower often much declined; branchlets many, the flat spray disposed in fan-shaped planes at different angles; foliage bright, often interspersed here and there with yellow, faded leaves.

Bark.—Bark of trunk in old trees a dead ash-gray, striate with broad and flat ridges, often conspicuously spirally twisted, shreddy at the edge; young stems and large branches reddish-brown, more or less striate and shreddy; branchlets ultimately smooth, shining, reddish-brown, marked by raised scars; season's twigs invested with leaves. [Pg 24]

Winter Buds and Leaves.—Leaf-buds naked, minute. Leaves in opposite pairs, 4-ranked, closely adherent to the branchlet and completely covering it, keeled in the side pairs and flat in the others, scale-like, ovate (in seedlings needle-shaped), obtuse or pointed at the apex, glandular upon the back, exhaling when bruised a strong aromatic odor

Inflorescence.—April to May. Flowers terminal, dark reddish-brown; sterile and fertile, usually on the same plant, rarely on separate plants; anthers opposite; filaments short; ovuliferous scales opposite, with slight projections near the base, usually 2-ovuled.

Fruit.—Cones, terminal on short branchlets, spreading or recurved, about ½ inch long, reddish-brown, loose-scaled, opening to the base at maturity; persistent through the first winter; scales 6-12, dry, oblong, not shield-shaped, not pointed; margin entire or nearly so; seeds winged all round.

Horticultural Value.—Hardy in New England; adapts itself to all soils and exposures, but prefers moist locations; grows slowly. Young trees have a narrowly conical outline, which spreads out at the base with age; retains its lower branches in open places, and is especially useful for hedges or narrow evergreen screens; little affected by insects; often disfigured, however, by dead branches and discolored leaves; is transplanted readily, and can be obtained in any quantity from nurserymen and collectors. The horticultural forms in cultivation range from thick, low, spreading tufts, through very dwarf, round, oval or conical forms, to tall, narrow, pyramidal varieties. Some have all the foliage tinged bright yellow, cream, or white; others have variegated foliage; another form has drooping branches. The bright summer foliage turns to a brownish color in winter. It is propagated from the seed and its horticultural forms from cuttings and layers.

Plate XI. — Thuja occidentalis.

1. Flowering branch with the preceding year's fruit.
2. Branch.
3. Sterile flower.
4. Stamen.
5. Fertile flower.
6. Scale with ovules.

[Pg 25]

Cupressus thyoides, L.

Chamæcyparis sphæroidea, Spach. Chamæcyparis thyoides, B. S. P.

White Cedar. Cedar.

Habitat and Range. — In deep swamps and marshes, which it often fills to the exclusion of other trees, mostly near the seacoast.

Cape Breton island and near Halifax, Nova Scotia, perhaps introduced in both.

Maine, — reported from the southern part of York county; New Hampshire, — limited to Rockingham county near the coast; Vermont, — no station known; Massachusetts, — occasional in central and eastern sections, very common in the southeast; Rhode Island, — common; Connecticut, — occasional in peat swamps.

Southward, coast region to Florida and west to Mississippi.

Habit. — 20-50 feet high and 1-2 feet in diameter at the ground, reaching in the southern states an altitude of 90 and a diameter of 4 feet; trunk straight, tapering slowly, throwing out nearly horizontal, slender branches, forming a narrow, conical head often of great elegance and lightness; foliage light brownish-green; strong-scented; spray flat in planes disposed at different angles; wood permanently aromatic.

Bark. — Bark of trunk thick, reddish, fibrous, shreddy, separating into thin scales, becoming more or less furrowed in old trees; branches reddish-brown; fine scaled; branches after fall of leaves, in the third or fourth year, smooth, purplish-brown; season's shoots at first greenish.

Winter Buds and Leaves. — Leaf-buds naked, minute. Leaves mostly opposite, 4-ranked, adherent to the branchlet and completely covering it; keeled in the side pairs and slightly convex in the others, dull green, pointed at apex or triangular awl-shaped, mostly with a minute roundish gland upon the back.

Inflorescence. — April. Flowers terminal, sterile and fertile, [Pg 26] usually on the same plant, rarely on separate plants, fertile on short branchlets: sterile, globular or oblong, anthers opposite, filaments shield-shaped: fertile, oblong or globular; ovuliferous scales opposite, slightly spreading at top, dark reddish-brown.

Fruit.—Cones, variously placed, ½ inch in diameter, roundish, purplish-brown, opening towards the center, never to the base; scales shield-shaped, woody; seeds several under each scale, winged.

Horticultural Value.—Hardy throughout New England, growing best in the southern sections. Young trees are graceful and attractive, but soon become thin and lose their lower branches; valued chiefly in landscape planting for covering low and boggy places where other trees do not succeed as well. Seldom for sale in nurseries, but easily procured from collectors. Several unimportant horticultural forms are grown.

Plate XII. — Cupressus thyoides.

1. Branch with flowers.
2. Sterile flower.
3. Stamen, back view.
4. Stamen, front view.
5. Fertile flower.
6. Ovuliferous scale with ovules.
7. Fruiting-branch.
8. Fruit.
9. Branch.

Juniperus Virginiana, L.

Red Cedar. Cedar. Savin.

Habitat and Range. — Dry, rocky hills but not at great altitudes, borders of lakes and streams, sterile plains, peaty swamps.

Nova Scotia and New Brunswick to Ontario.

Maine, — rare, though it extends northward to the middle Kennebec valley, reduced almost to a shrub; New Hampshire, — most frequent in the southeast part of the state; sparingly in the Connecticut valley as far north as Haverhill (Grafton [Pg 27] county); found also in Hart's location in the White mountain region; Vermont, — not abundant; occurs here and there on hills at levels less than 1000 feet; frequent in the Champlain and lower Connecticut valleys; Massachusetts, — west and center occasional, eastward common; Rhode Island and Connecticut, — common.

South to Florida; west to Dakota, Nebraska, Kansas, and Indian Territory.

Habit. — A medium-sized tree, 25-40 feet high, with a trunk diameter of 8-20 inches, attaining much greater dimensions southward; extremely variable in outline; the lower branches usually nearly horizontal, the upper ascending; head when young very regular, narrow-based, close and conical; in old trees frequently rather open, wide-spreading, ragged, roundish or flattened. In very exposed situations, especially along the seacoast, the trunk sometimes rises a foot or two and then develops horizontally, forming a curiously contorted lateral head. Under such conditions it occasionally becomes a dwarf tree 2-3 feet high, with wide-spreading branches and a very dense dome; spray close, foliage a sombre green, sometimes tinged with a rusty brownish-red; wood pale red, aromatic.

Bark. — Bark of trunk light reddish-brown, fibrous, shredding off, now and then, in long strips, exposing the smooth brown inner bark; season's shoots green.

Winter Buds and Leaves. — Leaf-buds naked, minute. Leaves dull green or brownish-red, of two kinds:

1. Scale-like, mostly opposite, each pair overlapping the pair above, 4-ranked, ovate, acute, sometimes bristle-tipped, more or less convex, obscurely glandular.

2. Scattered, not overlapping, narrowly lanceolate or needle-shaped, sharp-pointed, spreading. The second form is more common in young trees, sometimes comprising all the foliage, but is often found on trees of all ages, sometimes aggregated in dense masses.

Inflorescence. — Early May. Flowers terminating short branches, sterile and fertile, more commonly on separate trees, [Pg 28] often on the same tree; anthers in opposite pairs; ovuliferous scales in opposite pairs, slightly spreading, acute or obtuse; ovules 1-4.

Fruit. — Berry-like from the coalescence of the fleshy cone-scales, the extremities of which are often visible, roundish, the size of a small pea, dark blue beneath a whitish bloom, 1-4-seeded.

Horticultural Value. — Hardy throughout New England; prefers sunny slopes and a loamy soil, but grows well in poor, thin soils and upon wind-swept sites; young plants increase in height 1-2 feet yearly and have a very formal, symmetrical outline; old trees often become irregular and picturesque, and grow very slowly; a long-lived tree; usually obtainable in nurseries and from collectors, but must frequently be transplanted to be moved with safety. If a ball of earth can be retained about the roots of wild plants, they can often be moved successfully. There are horticultural forms distinguished by a slender weeping or distorted habit, and by variegated bluish or yellowish foliage, occasionally found in American nurseries. The type is usually propagated from the seed, the horticultural forms from cuttings or by grafting.

Plate XIII. — Juniperus Virginiana.

1. Branch with sterile and fertile flowers.
2. Sterile flower.
3. Stamen with pollen-sacs.
4. Fertile flower.
5. Fruiting branch.
6. Branch.
7. Branch with needle-shaped leaves.

SALICACEÆ. WILLOW FAMILY.

Trees or shrubs; leaves simple, alternate, undivided, with stipules either minute and soon falling or leafy and persistent; inflorescence from axillary buds of the preceding season, appearing with or before the leaves, in nearly erect, spreading or drooping catkins, sterile and fertile on separate [Pg 29] trees; flowers one to each bract, without calyx or corolla; stamens one to many; style short or none; stigmas 2, entire or 2-4-lobed; fruit a 2-4-celled capsule.

POPULUS.

Inflorescence usually appearing before the leaves; flowers with lacerate bracts, disk cup-shaped and oblique-edged, at least in sterile flowers; stamens usually many, filaments distinct; stigmas mostly divided, elongated or spreading.

SALIX.

Inflorescence appearing with or before the leaves; flowers with entire bracts and one or two small glands; disks wanting; stamens few.

Populus tremuloides, Michx.

Poplar. Aspen.

Habitat and Range.—In all soils and situations except in deep swamps, though more usual in dry uplands; sometimes springing up in great abundance in clearings or upon burnt lands.

Newfoundland, Labrador, and Nova Scotia to the Hudson bay region and Alaska.

New England,—common, reaching in the White mountain region an altitude of 3000 feet.

South to New Jersey, along the mountains in Pennsylvania and Kentucky, ascending 3000 feet in the Adirondacks; west to the slopes of the Rocky mountains, along which it extends to Mexico and Lower California.

Habit.—A graceful tree, ordinarily 35-40 feet and not uncommonly 50-60 feet high; trunk 8-15 inches in diameter, tapering, surmounted by a very open, irregular head of small, spreading branches; spray sparse, consisting of short, stout, leafy rounded shoots set at a wide angle; distinguished by the slenderness of its habit, the light color of trunk and [Pg 30] branches, the deep red of the sterile catkins in early spring, and the almost ceaseless flutter of the delicate foliage.

Bark. Trunk pale green, smooth, dark-blotched below the branches, becoming ash-gray and roughish in old trees; season's shoots dark reddish-brown or green, shining; bitter.

Winter Buds and Leaves.—Buds ⅛-¼ inch long, reddish-brown and lustrous, usually smooth, ovate, acute, often slightly incurved at apex, the upper often appressed. Leaves 1-2½ inches long, breadth usually equal to or exceeding the length, yellowish-green and ciliate when young, dark dull green above when mature, lighter beneath, glabrous on both sides, bright yellow in autumn; outline broadly ovate to orbicular, finely serrate or wavy-edged, with incurved, glandular-tipped teeth, apex rather abruptly acute or short-acuminate; base acute, truncate or slightly heart-shaped, 3-nerved; leafstalk slender, strongly flattened at right angles to the plane of the blade, bending to the slightest breath of air; stipules lanceolate, silky, soon falling.

Inflorescence.—April to May. Sterile catkins 1-3 inches long, fertile at first about the same length, gradually elongating; bracts cut into several lanceolate or linear divisions, silky-hairy; stamens about 10; anthers red: ovary short-stalked; stigmas two, 2-lobed, red.

Fruit.—June. Capsules, in elongated catkins, conical; seeds numerous, white-hairy.

Horticultural Value.—Hardy throughout New England in the most exposed situations; grows almost anywhere, but prefers a moist, rich loam; grows rapidly; foliage and spray thin; generally short-lived; often used as a screen for slow-growing trees; type seldom found in nurseries, but one or two horticultural forms are occasionally offered. Propagated from seed or cuttings.

Plate XIV. — Populus tremuloides.

1. Branch with sterile catkins.
2. Sterile flower.
3. Branch with fertile catkins.
4. Fertile flower.
5. Fruiting branch.
6. Branch with mature leaves.
7. Variant leaves.

[Pg 31]

Populus grandidentata, Michx.

Poplar. Large-toothed Aspen.

Habitat and Range. — In rich or poor soils; woods, hillsides, borders of streams.

Nova Scotia, New Brunswick, southern Quebec, and Ontario.

New England, — common, occasional at altitudes of 2000 feet or more.

South to Pennsylvania and Delaware, along the mountains to Kentucky, North Carolina, and Tennessee; west to Minnesota.

Habit. — A tree 30-45 feet in height and 1 foot to 20 inches in diameter at the ground, sometimes attaining much greater dimensions; trunk erect, with an open, unsymmetrical, straggling head; branches distant, small and crooked; branchlets round; spray sparse, consisting of short, stout, leafy shoots; in time and manner of blossoming, constant motion of foliage, and general habit, closely resembling *P. tremuloides*.

Bark. — Bark of trunk on old trees dark grayish-brown or blackish, irregularly furrowed, broad-ridged, the outer portions separated into small, thickish scales; trunk of young trees soft greenish-gray; branches greenish-gray, darker on the underside; branchlets dark greenish-gray, roughened with leaf-scars; season's twigs in fall dark reddish-brown, at first tomentose, becoming smooth and shining.

Winter Buds and Leaves. — Buds ⅛ inch long, mostly divergent, light chestnut, more or less pubescent, dusty-looking, ovate, acute. Leaves 3-5 inches long, two-thirds as wide, densely white-tomentose when opening, usually smooth on both sides when mature, dark green above, lighter beneath, bright yellow in autumn; outline roundish-ovate, coarsely and irregularly sinuate-toothed; teeth acutish; sinuses in shallow curves; apex acute; base truncate or slightly heart-shaped; leafstalks long, strongly flattened at right angles to the plane of the blade; stipules thread-like, soon falling.

Inflorescence. — March to April. Sterile catkins 1-3 inches long, fertile at first about the same length, but gradually [Pg 32] elongating; bracts cut into several lanceolate divisions, silky-hairy; stamens about 10; anthers red: ovaries short-stalked; stigmas two, 2-lobed, red.

Fruit. — Fruiting catkins at length 3-6 inches long; capsule conical, acute, roughish-scurfy, hairy at tip: seeds numerous, hairy.

Horticultural Value. — Hardy throughout New England; grows almost anywhere, but prefers moist, rich loam; grows rapidly and is safely transplanted, but is unsymmetrical, easily broken by the wind, and short-lived; seldom offered by nurserymen, but readily procured from northern collectors of native plants. Useful to grow for temporary effect with permanent trees, as it will fail by the time the desirable kinds are well established. Propagated from seed or cuttings.

Note. — Points of difference between *P. tremuloides* and *P. grandidentata*. These trees may be best distinguished in early spring by the color of the unfolding leaves. In the sunlight the head of *P. tremuloides* appears yellowish-green, while that of *P. grandidentata* is conspicuously cotton white. The leaves of *P. grandidentata* are larger and more coarsely toothed, and the main branches go off usually at a broader angle. The buds of *P. grandidentata* are mostly divergent, dusty-looking, dull; of *P. tremuloides*, mostly appressed, highly polished with a resinous lustre.

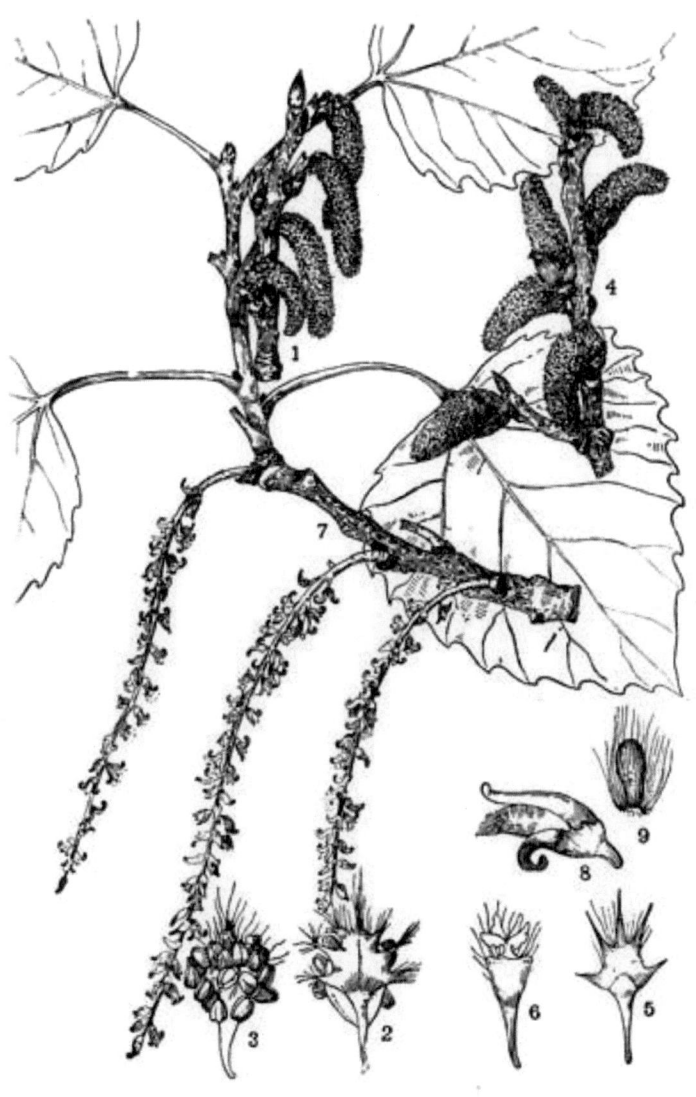

Plate XV. — Populus grandidentata.

1. Branch with sterile catkins.
2. Sterile flower, back view,
3. Sterile flower, front view.
4. Branch with fertile catkins.
5. Bract of fertile flower.
6. Fertile flower, front view.
7. Fruiting branch with mature leaves.
8. Fruit.
9. Fruit.

[Pg 33]

Populus heterophylla, L.

Poplar. Swamp Poplar. Cottonwood.

Habitat and Range. — In or along swamps occasionally or often overflowed; rare, local, and erratically distributed.

Connecticut, — frequent in the southern sections; Bozrah (J. N. Bishop); Guilford, in at least three wood-ponds (W. E. Dudley *in lit.*), New Haven, and near Norwich (W. A. Setchell).

Following the eastern coast in wide belts from New York (Staten island and Long island) south to Georgia; west along the Gulf coast to western Louisiana, and northward along the Mississippi and Ohio basins to Arkansas, Indiana, and Illinois.

Habit. — A slender, medium-sized tree, attaining a height of 30-50 feet, reaching farther south a maximum of 90 feet; trunk 9-18 inches in diameter, usually branching high up, forming a rather open hemispherical or narrow-oblong head; branches irregular, short, rising, except the lower, at a sharp angle; branchlets stout, roundish, varying in color, degree of pubescence, and glossiness, becoming rough after the first year with the raised leaf-scars; spray sparse.

Bark. — Bark of trunk dark ash-gray, very rough, and broken into loosely attached narrow plates in old trees; in young trees light ash-gray, smooth at first, becoming in a few years roughish, low-ridged.

Winter Buds and Leaves. — Buds conical, acute, more or less resinous. Leaves 3-6 inches long, two-thirds as wide, densely white-tomentose when young, at length dark green on the upper side, lighter beneath and smooth except along the veins; outline ovate, wavy-toothed; base heart-shaped, lobes often overlapping; apex obtuse; leafstalk long, round, downy; stipules soon falling.

Inflorescence. — April to May. Sterile catkins when expanded 3-4 inches long, at length pendent; scales cut into irregular divisions, reddish; stamens numerous, anthers oblong, dark red: fertile catkins spreading, few and loosely flowered, gradually [Pg 34] elongating; scales reddish-brown; ovary short-stalked; styles 2-3, united at the base; stigmas 2-3, conspicuous.

Fruit. — Fruiting catkins spreading or drooping, 4-5 inches long: capsules usually erect, ovoid, acute, shorter than or equaling the slender pedicels: seeds numerous, white-hairy.

Horticultural Value. — Not procurable in New England nurseries or from collectors; its usefulness in landscape gardening not definitely known.

Plate XVI. — Populus heterophylla.

1. Winter buds.
2. Branch with sterile catkin.
3. Sterile flower.
4. Scale of sterile flower.
5. Branch with fertile catkin.
6. Fertile flower.
7. Fruiting branch with mature leaves.

Populus deltoides, Marsh.

Populus monilifera, Ait.

Cottonwood. Poplar.

Habitat and Range. — In moist soil; river banks and basins, shores of lakes, not uncommon in drier locations.

Throughout Quebec and Ontario to the base of the Rocky mountains.

Maine, — not reported; New Hampshire, — restricted to the immediate vicinity of the Connecticut river, disappearing near the northern part of Westmoreland; Vermont, — western sections, abundant along the shores of the Hoosac river in Pownal and along Lake Champlain (W. W. Eggleston); in the Connecticut valley as far north as Brattleboro (*Flora of Vermont*, 1900); Massachusetts, — along the Connecticut and its tributaries; Rhode Island, — occasional; Connecticut, — occasional eastward, common along the Connecticut, Farmington, and Housatonic rivers.

South to Florida; west to the Rocky mountains.

Habit. — A stately tree, 75-100 feet in height; trunk 3-5 feet in diameter, light gray, straight or sometimes slightly inclined, of nearly uniform size to the point of branching, [Pg 35] surmounted by a noble, broad-spreading, open, symmetrical head, the lower branches massive, horizontal, or slightly ascending, more or less pendulous at the extremities, the upper coarse and spreading, rising at a sharper angle; branchlets stout; foliage brilliant green, easily set in motion; the sterile trees gorgeous in spring with dark red pendent catkins.

Bark. — In old trees thick, ash-gray, separated into deep, straight furrows with rounded ridges; in young trees light yellowish-green, smooth; season's shoots greenish, marked with pale longitudinal lines.

Winter Buds and Leaves. — Buds large, conical, smooth, shining. Leaves 3-6 inches long, scarcely less in width, variable in color and shape, ordinarily dark green and shining above, lighter beneath, ribs raised on both sides; outline broadly ovate, irregularly crenate-toothed; apex abruptly acute or acuminate; base truncate, slightly heart-shaped or sometimes acute; stems long, slender, somewhat flattened at right angles to the plane of the blade; stipules linear, soon falling.

Inflorescence. — April to May. In solitary, densely flowered catkins; bracts lacerate-fringed, each bract subtending a cup-shaped scale; stamens very numerous; anthers longer than the filaments, dark red: fertile catkins elongating to 5 or 6 inches; ovary ovoid; stigmas 3 or 4, nearly sessile, spreading.

Fruit. — Capsules ovate, rough, short-stalked; seeds densely cottony.

Horticultural Value. — Hardy in southern-central New England; grows rapidly in almost any soil and is readily obtainable in nurseries. Where an immediate effect is desired, the cottonwood serves the purpose excellently and frequently makes very fine large individual trees, but the wood is soft and likely to be broken by wind or ice. Usually propagated from cuttings.

Plate XVII.—Populus deltoides.

1. Winter buds.
2. Branch with sterile catkins.
3. Sterile flower, back view.
4. Sterile flower, front view.
5. Scale of sterile flower.
6. Fertile flower.
7. Fruiting catkin.
8. Branch with mature leaves.
9. Variant leaf.

[Pg 36]

Populus balsamifera, L.

Balsam. Poplar. Balm of Gilead.

Habitat and Range.—Alluvial soils; river banks, valleys, borders of swamps, woods.

Newfoundland and Nova Scotia west to Manitoba; northward to the coast of Alaska and along the Mackenzie river to the Arctic circle.

Maine,—common; New Hampshire,—Connecticut river valley, generally near the river, becoming more plentiful northward; Vermont,—frequent; Massachusetts and Rhode Island,—not reported; Connecticut,—extending along the Housatonic river at New Milford for five or six miles, perhaps derived from an introduced tree (C. K. Averill, *Rhodora*, II, 35).

West through northern New York, Michigan, Minnesota, Dakota (Black Hills), Montana, beyond the Rockies to the Pacific coast.

Habit.—A medium-sized tree, 30-75 feet high, trunk 1-3 feet in diameter, straight; branches horizontal or nearly so, slender for size of tree, short; head open, narrow-oblong or oblong-conical; branchlets mostly terete; foliage thin.

Bark.—In old trees dark gray or ash-gray, firm-ridged, in young trees smooth; branchlets grayish; season's shoots reddish or greenish brown, sparsely orange-dotted.

Winter Buds and Leaves.—Buds 3/4 inch long, appressed or slightly divergent, conical, slender, acute, resin-coated, sticky, fragrant when opening. Leaves 3-6 inches long, about one-half as wide, yellowish when young, when mature bright green, whitish below; outline ovate-lanceolate or ovate, finely toothed, gradually tapering to an acute or acuminate apex; base obtuse to rounded, sometimes truncate or heart-shaped; leafstalk much shorter than the blade, terete or nearly so; stipules soon falling. The leaves of var. *intermedia* are obovate to oval; those of var. *latifolia* closely approach the leaves of *P. candicans*.

Inflorescence.—April. Sterile 3-4 inches long, fertile at first about the same length, gradually elongating, loosely flowered; [Pg 37] bracts irregularly and rather narrowly cut-toothed, each bract subtending a cup-shaped disk; stamens numerous; anthers red: ovary short-stalked; stigmas two, 2-lobed, large, wavy-margined.

Fruit.—Fruiting catkins drooping, 4-6 inches long: capsules ovoid, acute, longer than the pedicels, green: seeds numerous, hairy.

Horticultural Value.—Hardy throughout New England; grows in all excepting very wet soils, in full sun or light shade, and in exposed situations; of rapid growth, but subject to the attacks of borers, which kill the branches and make the head unsightly; also spreads from the roots, and therefore not desirable for ornamental plantations; most useful in the formation of shelter-belts; readily transplanted but not common in nurseries. Propagated from cuttings.

Plate XVIII.—Populus balsamifera.

1. Branch with sterile flowers.
2. Sterile flower, back view.
3. Sterile flower, side view.
4. Scales of sterile flower.
5. Branch with fertile catkins.
6. Fertile flower.
7. Fruiting catkins, mature.
8. Branch with mature leaves.

Populus candicans, Ait.

Populus balsamifera, **var.** *candicans, Gray.*

Balm of Gilead.

Habitat and Range. — In a great variety of soils; usually in cultivated or pasture lands in the vicinity of dwellings; infrequently found in a wild state. The original site of this tree has not been definitely agreed upon. Professor L. H. Bailey reports that it is indigenous in Michigan, and northern collectors find both sexes in New Hampshire and Vermont; while in central and southern New England the staminate tree is rarely if ever seen, and the pistillate flowers seldom [Pg 38] if ever mature perfect fruit. The evidence seems to indicate a narrow belt extending through northern New Hampshire, Vermont and Michigan, with the intermediate southern sections of the Province of Ontario as the home of the Balm of Gilead.

Nova Scotia and New Brunswick, — occasional; Ontario, — frequent.

New England, — occasional throughout.

South to New Jersey; west to Michigan and Minnesota.

Habit. — A medium-sized tree, 40-60 feet high; trunk 1-3 feet in diameter, straight or inclined, sometimes beset with a few crooked, bushy branchlets; head very variable in shape and size; solitary in open ground, commonly *broad-based, spacious, and pyramidal,* among other trees more often rather small; loosely and irregularly branched, with sparse, coarse, and often crooked spray; *foliage dark green, handsome, and abundant;* all parts characterized by a strong and peculiar resinous fragrance. A single tree multiplying by suckers often becomes parent of a grove covering half an acre, more or less, made up of trees of all ages and sizes.

Bark. — Bark of trunk and lower portions of large branches dark gray, rough, irregularly striate and firm in old trees; in young trees and upon smaller branches smooth, soft grayish-green, often flanged by prominent ridges running down the stalk from the vertices of the triangular leaf-scars; season's shoots often flanged, shining reddish or olive green, with occasional longitudinal gray lines, viscid.

Winter Buds and Leaves.—Buds dark reddish-brown, rather closely set along the stalk, conical or somewhat angled, narrow, often falcate, sharp-pointed, resinous throughout, viscid, aromatic, exhaling a powerful odor when the scales expand, terminal about 3/4 inch long. Leaves 4-6 inches long and nearly as wide, yellowish-green at first, becoming dark green and smooth on the upper surface with the exception of a *minute pubescence along the veins*, dull light green beneath, finely serrate with incurved glandular points, usually ciliate with minute stiff, whitish hairs; base heart-shaped; apex short-pointed; [Pg 39] petioles about 1-1½ inches long, *more or less hairy*, somewhat flattened at right angles to the blade; stipules short, ovate, acute, soon falling.

Inflorescence.—Similar to that of *P. balsamifera*.

Fruit.—Similar to that of *P. balsamifera*.

Horticultural Value.—Hardy throughout New England; has an attractive foliage and grows rapidly in all soils and situations, but the branches are easily broken by the wind, and its habit of suckering makes it objectionable in ornamental ground; occasionally offered by nurserymen and collectors. Propagated from cuttings.

119

Plate XIX. — Populus candicans.

1. Winter bud.
2. Branch with fertile catkins.
3. Fertile flower.
4. Fruiting branch.

Populus alba, L.

Abele. White Poplar. Silver-leaf Poplar.

Range. — Widely distributed in the Old World, extending in Europe from southern Sweden to the Mediterranean, throughout northern Africa, and eastward in Asia to the northwestern Himalayas. Introduced from England by the early settlers and soon established in the colonial towns, as in Plymouth and Duxbury, on the western shore of Massachusetts bay. Planted or spontaneous over a wide area.

New Brunswick and Nova Scotia, — occasional.

New England, — occasional throughout, local, sometimes common.

Southward to Virginia.

Habit. — A handsome tree, resembling *P. grandidentata* more than any other American poplar, but of far nobler proportions; 40-75 feet high and 2-4 feet in diameter at the ground; growing much larger in England; head large, spreading; round-topped, in spring enveloped in a dazzling [Pg 40] cloud of cotton white, which resolves itself later into two conspicuously contrasting surfaces of dark green and silvery white.

Bark. — Light gray, smooth upon young trees, in old trees furrowed upon the trunk.

Winter Buds and Leaves. — Buds not viscid, cottony. Leaves 1-4 inches long, densely white-tomentose while expanding, when mature dark green above and white-tomentose to glabrous beneath;

outline ovate or deltoid, 3-5-lobed and toothed or simply toothed, teeth irregular; base heart-shaped or truncate; apex acute to obtuse; leafstalk long, slender, compressed; stipules soon falling.

Inflorescence and Fruit. — April to May. Sterile catkins 2-4 inches long, cylindrical, fertile at first shorter, — stamens 6-16; anthers purple: capsules ¼ inch long, narrow-ovoid; seeds hairy.

Horticultural Value. — Hardy. Thrives even in very poor soils and in exposed situations; grows rapidly in good soils; of distinctive value in landscape gardening but not adapted for planting along streets and upon lawns of limited area on account of its habit of throwing out numerous suckers and its liability to damage from heavy winds. The sides of country roads where the abele has been planted are sometimes obstructed for a considerable distance by the thrifty shoots from underground.

Salix discolor. Muhl.

Pussy Willow. Glaucous Willow.

Habitat and Range. — Low, wet grounds; banks of streams, swamps, moist hillsides.

Nova Scotia to Manitoba.

Maine, — abundant; common throughout the other New England states.

South to North Carolina; west to Illinois and Missouri.

Habit. — Mostly a tall shrub with several stems, but occasionally assuming a tree-like habit, with a height of 15-20 feet [Pg 41] and trunk diameter of 5-10 inches; one tree reported at Laconia, N. H., 35 feet high (F. W. Batchelder); branches few, stout, ascending, forming a very open, hemispherical head.

Bark. — Trunk reddish-brown; branches dark-colored; branchlets light green, orange-dotted.

Winter Buds and Leaves. — Buds ovate-conical; apex obtuse to acute. Leaves simple, alternate, 2-4 inches long, smooth and bright green above, smooth and whitish beneath when fully grown; outline ovate-lanceolate to narrowly oblong-oval, crenulate-serrate to entire; apex acute, base acute and entire; leafstalk short; stipules toothed or entire.

Inflorescence. — March to April. Appearing before the leaves in catkins, sterile and fertile on separate plants, occasionally both kinds on the same plant, sessile, — sterile spreading or erect, oblong-cylindrical, silky; calyx none; petals none; bracts entire, reddish-brown turning to black, oblong to oblong-obovate, with long, silky hairs; stamens 2; filaments distinct: fertile catkins spreading; bracts oblong to ovate, hairy; style short; stigma deeply 4-lobed.

Fruit. — Fruiting catkins somewhat declined: capsules ovate-conical, tomentose, stem two-thirds the length of the scale: seeds numerous.

Horticultural Value. — Picturesque in blossom and fruit; its value dependent chiefly upon its matted roots for holding wet banks, and its ability to withstand considerable shade. Sold by plant collectors; easily propagated from cuttings.

Plate XX. — Salix discolor.

1. Leaf-buds.
2. Branch with sterile catkins.
3. Sterile flower.
4. Branch with fertile catkins.
5. Fertile flower.
6. Fruiting branch.
7. Mature leaves.

[Pg 42]

Salix nigra, Marsh.

Black Willow

Habitat and Range. — In low grounds, along streams or ponds, river flats.

New Brunswick to western Ontario.

New England, — occasional throughout, frequent along the larger streams.

South to Florida; west to Dakota, Nebraska, Kansas, Indian territory, Louisiana, Texas, southern California, and south into Mexico.

Habit. — A large shrub or small tree, 25-40 feet high and 10-15 inches in trunk diameter, attaining great size in the Ohio and Mississippi valleys and the valley of the lower Colorado; trunk short, surmounted by an irregular, open, often roundish head, with stout, spreading branches, slender branchlets, and twigs brittle towards their base.

S. nigra, var. *falcata*, Pursh., covers about the same range as the type and differs chiefly in its narrower, falcate leaves.

Bark. — Trunk rough, in young trees light brown, in old trees dark-colored or nearly black, deeply and irregularly ridged, separated on the surface into thick, plate-like scales; branchlets reddish-brown; twigs bronze olive.

Winter Buds and Leaves. — Buds narrowly conical, acute. Leaves simple, alternate, appearing much later than those of *S. discolor*, 2-5 inches long, somewhat pubescent on both sides when young, when mature green and smooth above, paler and sometimes pubescent along the veins beneath; outline narrowly lanceolate, finely serrate; apex acute or acuminate, often curved; base acutish to rounded or slightly heart-shaped; petiole short, usually pubescent; stipules large and persistent, or small and soon falling.

Inflorescence. — April to May. Appearing with the leaves from the axils of the short, lateral shoots, in catkins, sterile and fertile on different trees, stalked, — sterile spreading, narrowly cylindrical; calyx none; corolla none; bracts [Pg 43] entire, rounded to oblong, villous, ciliate; stamens about 5: fertile catkins spreading; calyx none; corolla none; bracts ovate to narrowly oblong, acute, villous; ovary short-stalked, with two small glands at its base, ovate-conical, sometimes obovate, smooth; stigmas 2, short.

Fruit. — Fertile catkins drooping: capsules ovate-conical, short-stemmed, minutely granular; style very short: seeds numerous.

Horticultural Value. — Hardy in New England, grows rapidly in all soils, particularly useful in very wet situations; seriously affected by insects; occasionally offered in nurseries; transplanted readily; propagated from cuttings.

127

Plate XXI.—Salix nigra.

1. Winter buds.
2. Branch with sterile catkins.
3. Sterile flower, side view.
4. Sterile flower, front view.
5. Branch with fertile catkins.
6. Fertile flower, side view.
7. Fertile flower, front view.
8. Fruiting branch.
9. Fruit enlarged.

Salix fragilis and Salix alba.

The *fragilis* and *alba* group of genus *Salix* gives rise to puzzling questions of determination and nomenclature. Pure *fragilis* and pure *alba* are perfectly distinct plants, *fragilis* occasional, locally rather common, and *alba* rather rare within the limits of the United States. Each species has varieties; the two species hybridize with each other and with native species, and the hybrids themselves have varietal forms. This group affords a tempting field for the manufacture of species and varieties, about most of which so little is known that any attempt to assign a definite range would be necessarily imperfect and misleading. The range as given below in either species simply points out the limits within which any one of the various forms of that species appears to be spontaneous. [Pg 44]

Salix fragilis, L.

Crack Willow. Brittle Willow.

Habitat and Range.—In low land and along river banks. Indigenous in southwestern Asia, and in Europe where it is extensively cultivated; introduced into America probably from England for use in basket-making, and planted at a very early date in many of the colonial towns; now extensively cultivated, and often spontaneous in wet places and along river banks, throughout New England and as far south as Delaware.

Habit.—Tree often of great size; attaining a maximum height of 60-90 feet; head open, wide-spreading; branches except the lowest rising at a broad angle; branchlets reddish or yellowish green, smooth and polished, very brittle at the base. In 1890 there was standing upon the Groome estate, Humphreys Street, Dorchester, Mass., a willow of this species about 60 feet high, 28 feet 2 inches in girth five feet from the ground, with a spread of 110 feet (*Typical Elms and other Trees of Massachusetts*, p. 85).

Bark.—Bark of the trunk gray, smooth in young trees, in old trees very rough, irregularly ridged, sometimes cleaving off in large plates.

Winter Buds and Leaves.—Buds about ⅓ inch long, reddish-brown, narrow-conical. Leaves simple, alternate, 2-6 inches long, smooth, dark green and shining above, pale or glaucous beneath and somewhat pubescent when young; outline lanceolate, glandular-serrate; apex long-acuminate; tapering to an acute or obtuse base; leafstalk short, glandular at the top; stipules half-cordate when present, soon falling.

Inflorescence.—April to May. Catkins appearing with the leaves, spreading, stalked,—sterile 1-2 inches long; stamens 2-4, usually 2; filaments distinct, pubescent below; ovary abortive: fertile catkins slender; stigma nearly sessile; capsule long-conical, smooth, short-stalked.

Horticultural Value.—Hardy throughout New England; grows best near streams, but adapts itself readily to all rich, [Pg 45] damp soils. A handsome ornamental tree when planted where its roots can find water, and its branches space for free development. Readily propagated from slips.

Salix alba, L.

White Willow.

Habitat and Range. — Low, moist grounds; along streams. Probably indigenous throughout Europe, northern Africa, and Asia as far south as northwestern India. Extensively introduced in America, and often spontaneous over large areas.

New Brunswick, Nova Scotia, and Ontario.

New England, — sparingly throughout.

South to Delaware; extensively introduced in the western states.

Habit. — A large tree, 50-80 feet in height; trunk usually rather short and 2-7 feet in diameter; head large, not as broad-spreading as that of *S. fragilis*; branches numerous, mostly ascending.

Bark. — Bark of trunk in old trees gray and coarsely ridged, in young trees smooth; twigs smooth, olive.

Leaves. — Leaves simple, alternate, 2-4 inches long, *silky-hairy on both sides when young, when old still retaining more or less pubescence, especially on the paler under surface*; outline narrowly lanceolate or elliptic-lanceolate, glandular-serrate, tapering to a long pointed apex and to an acute base; leafstalk short, usually without glands; stipules ovate-lanceolate, soon falling.

Note. — Var. *vitellina*, Koch., by far the most common form of this willow; mature leaves glabrous above; twigs *yellow*. Var. *cærulea*, Koch.; mature leaves bluish-green, glabrous above, glaucous beneath; twigs *olive*.

Inflorescence. — April to May. Catkins appearing with the leaves, slender, erect, stalked; scales linear; stamens 2; filaments distinct, hairy below the middle; stigma nearly sessile, deeply cleft; capsule glabrous, sessile or nearly so. [Pg 46]

Horticultural Value. — Hardy throughout New England; grows best in moist localities; extensively cultivated to bind the soil along the banks of streams. Easily propagated from slips.

JUGLANDACEÆ. WALNUT FAMILY.

Juglans cinerea, L.

Butternut. Oilnut. Lemon Walnut.

Habitat and Range. — Roadsides, rich woods, river valleys, fertile, moist hillsides, high up on mountain slopes.

New Brunswick, throughout Quebec and eastern Ontario.

Maine, — common, often abundant; New Hampshire, — throughout the Connecticut valley, and along the Merrimac and its tributaries, to the base of the White mountains; Vermont, — frequent; Massachusetts, — common in the eastern and central portions, frequent westward; Rhode Island and Connecticut, — common.

South to Delaware, along the mountains to Georgia and Alabama; west to Minnesota, Kansas, and Arkansas.

Habit. — Usually a medium-sized tree, 20-45 feet in height, with a disproportionately large trunk, 1-4 feet in diameter; often attaining under favorable conditions much greater dimensions. It ramifies at a few feet from the ground and throws out long, rather stout, and nearly horizontal branches, the lower slightly drooping, forming for the height of the tree a very wide-spreading head, with a stout and stiffish spray. At its best the butternut is a picturesque and even beautiful tree.

Bark. — Bark of trunk dark gray, rough, narrow-ridged and wide-furrowed in old trees, in young trees smooth, dark gray; branchlets brown gray, with gray dots and prominent leaf-scars; season's shoots greenish-gray, faint-dotted, with a clammy pubescence. The bruised bark of the nut stains the skin yellow.

Winter Buds and Leaves. — Buds flattish or oblong-conical, few-scaled, 2-4 buds often superposed, the uppermost largest [Pg 47] and far above the axil. Leaves pinnately compound, alternate, 1-1½ feet long, viscid-pubescent throughout, at least when young; rachis enlarged at base; stipules none; leaflets 9-17, 2-4 inches long, about half as wide, upper surface rough, yellowish when unfolding in spring, becoming a dark green, lighter beneath, yellow in autumn; outline oblong-lanceolate, serrate; veins prominent beneath; apex acute to acuminate; base obtuse to rounded, somewhat inequilateral, sessile, except the terminal leaflet; stipels none.

Inflorescence. — May. Appearing while the leaves are unfolding, sterile and fertile flowers on the same tree, — the sterile from terminal or lateral buds of the preceding season, in single, unbranched, stout, green, cylindrical, drooping catkins 3-6 inches long; calyx irregular, mostly 6-lobed, borne on an oblong scale; corolla none; stamens 8-12, with brown anthers: fertile flowers sessile, solitary, or several on a common peduncle from the season's shoots; calyx hairy, 4-lobed, with 4 small petals at the sinuses; styles 2, short; stigmas 2, large, feathery, diverging, rose red.

Fruit. — Ripening in October, one or several from the same footstalk, about 3 inches long, oblong, pointed, green, downy, and sticky at first, dark brown when dry: shells sculptured, rough: kernel edible, sweet but oily.

Horticultural Value. — Hardy throughout New England; grows in any well-drained soil, but prefers a deep, rich loam; seldom reaches its best under cultivation. Trees of the same age are apt to vary in vigor and size, dead branches are likely to appear early, and sound trees 8 or 10 inches in diameter are seldom seen; the foliage is thin, appears late and drops early; planted in private grounds chiefly for its fruit; only occasionally offered in nurseries, collected plants seldom successful. Best grown from seed planted where the tree is to stand, as is evident from many trees growing spontaneously.

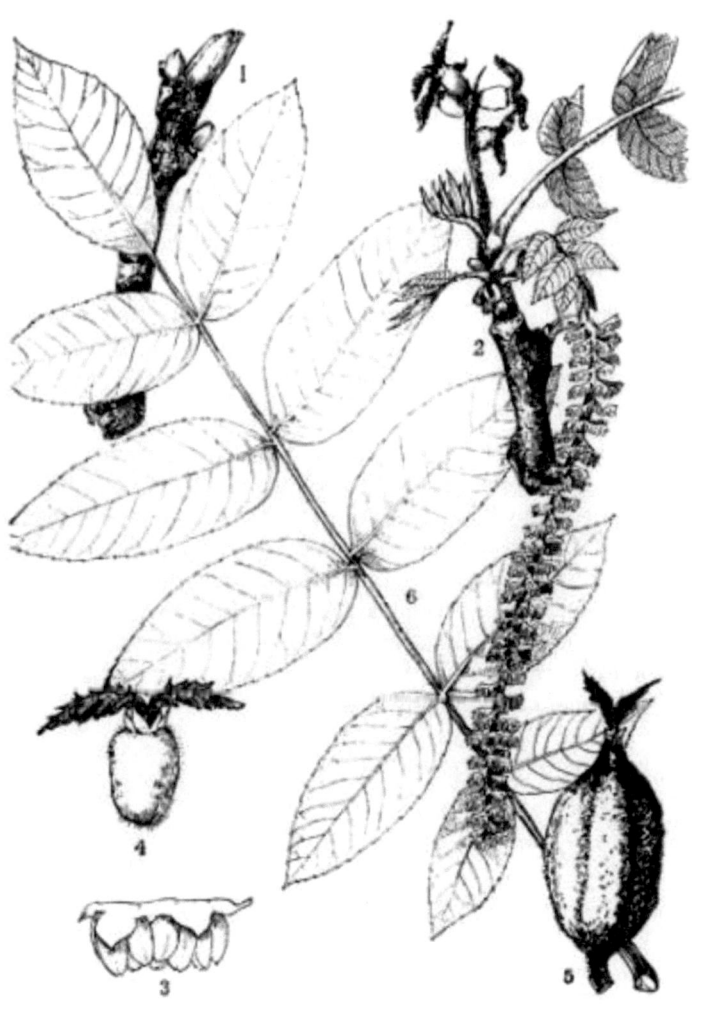

Plate XXII. — Juglans cinerea.

1. Winter buds.
2. Flowering branch.
3. Sterile flower, side view.
4. Fertile flower.
5. Fruit.
6. Leaf.

[Pg 48]

Juglans nigra, L.

Black Walnut.

Habitat and Range. — Rich woods.

Maine, New Hampshire, and Vermont, — not reported native; Massachusetts, — rare east of the Connecticut river, occasional along the western part of the Connecticut valley to the New York line; Rhode Island, — doubtfully native, Apponaug (Kent county) and elsewhere; Connecticut, — frequent westward, Darien (Fairfield county); Plainville (Hartford county, J. N. Bishop *in lit.*, 1896); in the central and eastern sections probably introduced.

South to Florida; west to Minnesota, Kansas, Arkansas, and Texas.

Habit. — A large tree, 50-75 feet high, with a diameter above the swell of the roots of 2-5 feet; attaining in the Ohio valley a height of 150 feet and a diameter of 6-8 feet; trunk straight, slowly tapering, throwing out its lower branches nearly horizontally, the upper at a broad angle, forming an open, spacious, noble head.

Bark. — Bark of trunk in old trees thick, blackish, and deeply furrowed; large branches rough and more or less furrowed; branchlets smooth; season's twigs downy.

Winter Buds and Leaves. — Buds small, ovate or rounded, obtuse, more or less pubescent, few-scaled. Leaves pinnately compound, alternate; rachis smooth and swollen at base, but less so than that of the butternut; stipules none; leaflets 13-21 (the odd leaflet at the apex often wanting), opposite or alternate, 2-5 inches long, about half as wide; dark green and smooth above, lighter and slightly glandular-pubescent beneath, turning yellow in autumn; outline ovate-lanceolate; apex taper-pointed; base oblique, usually rounded or heart-shaped; stemless or nearly so, except the terminal leaflet; stipels none. Aromatic when bruised.

Inflorescence. — May. Appearing while the leaves are unfolding, sterile and fertile flowers on the same tree, — the sterile along the sides or at the ends of the preceding year's branches, [Pg 49] in single, unbranched, green, stout, cylindrical, pendulous catkins, 3-6 inches long; perianth of 6 rounded lobes, stamens numerous, filaments very short, anthers purple: fertile flowers in the axils of the season's shoots, sessile, solitary or several on a common peduncle; calyx 4-toothed, with 4 small petals at the sinuses; stigmas 2, reddish-green.

Fruit. — Ripening in October at the ends of the branchlets, single, or two or more together; round, smooth, or somewhat roughish with uneven surface, not viscid, dull green turning to brown: husk not separating into sections: shell irregularly furrowed: kernel edible.

Horticultural Value. — Hardy in central and southern New England; grows well in most situations, but in a deep rich soil it forms a large and handsome tree. Readily obtainable in western nurseries; transplants rather poorly, and collected plants are of little value. Its leaves appear late and drop early, and the fruit is often abundant. These disadvantages make it objectionable in many cases. Grown from seed.

Plate XXIII. — Juglans nigra.

1. Winter buds.
2. Flowering branch.
3. Sterile flower, front view.
4. Sterile flower, back view.
5. Fertile flower.
6. Fruiting branch.

Carya alba, Nutt.

Hicoria ovata, Britton.

Shagbark. Shagbark or Shellbark Hickory. Walnut.

Habitat and Range.—In various soils and situations, fertile slopes, brooksides, rocky hills.

Valley of the St. Lawrence.

Maine,—along or near the coast as far north as Harpswell (Cumberland county); New Hampshire,—common as far north as Lake Winnepesaukee; Vermont,—occasional along the Connecticut to Windsor, rather common in the Champlain [Pg 50] valley and along the western slopes of the Green mountains; Massachusetts, Rhode Island, and Connecticut,—common.

South to Delaware and along the mountains to Florida; west to Minnesota, Kansas, Indian territory, and Texas.

Habit.—The tallest of the hickories and proportionally the most slender, from 50 to 75 feet in height, and not more than 2 feet in trunk diameter; rising to a great height in the Ohio and Indiana river bottoms. The trunk, shaggy in old trees, rises with nearly uniform diameter to the point of furcation, throwing out rather small branches of unequal length and irregularly disposed, forming an oblong or rounded head with frequent gaps in the continuity of the foliage.

Bark.—Trunk in young trees and in the smaller branches ash-gray, smoothish to seamy; in old trees, extremely characteristic, usually shaggy, the outer layers separating into long, narrow, unequal plates, free at one or both ends, easily detachable; branchlets smooth and gray, with conspicuous leaf-scars; season's shoots stout, more or less downy, numerous-dotted.

Winter Buds and Leaves.—Buds tomentose, ovate to oblong, terminal buds large, much swollen before expanding; inner scales numerous, purplish-fringed, downy, enlarging to 5-6 inches in length as the leaves unfold. Leaves pinnately compound, alternate, 12-20 inches long; petiole short, rough, and somewhat swollen at base; stipules none; leaflets usually 5, sometimes 3 or 7, 3-7 inches long, dark green above, yellowish-green and downy beneath when young, the three upper large, obovate to lanceolate, the two lower much smaller, oblong to oblong-lanceolate, all finely serrate and sharp-pointed; base obtuse, rounded or acute, mostly inequilateral; nearly sessile save the odd leaflet; stipels none.

Inflorescence. — May. Sterile and fertile flowers on the same tree, appearing when the leaves are fully grown, — sterile at the base of the season's shoots, in slender, green, pendulous catkins, 4-6 inches long, usually in threes, branching umbel-like from a common peduncle; flower-scales 3-parted, the middle lobe much longer than the other two, linear, tipped with long [Pg 51] bristles; calyx adnate to scale; stamens mostly in fours, anthers yellow, bearded at the tip: fertile flowers single or clustered on peduncles at the ends of the season's shoots; calyx 4-toothed, hairy, adherent to ovary; corolla none; stigmas 2, large, fringed.

Fruit. — October. Spherical, 3-6 inches in circumference: husks rather thin, firm, green turning to brown, separating completely into 4 sections: nut variable in size, subglobose, white, usually 4-angled: kernel large, sweet, edible.

Horticultural Value. — Hardy throughout New England; prefers light, well-drained, loamy soil; when well established makes a moderately rapid growth; difficult to transplant, rarely offered in nurseries; collected plants seldom survive; a fine tree for landscape gardening, but its nuts are apt to make trouble in public grounds. Propagated from a seed. A thin-shelled variety is in cultivation.

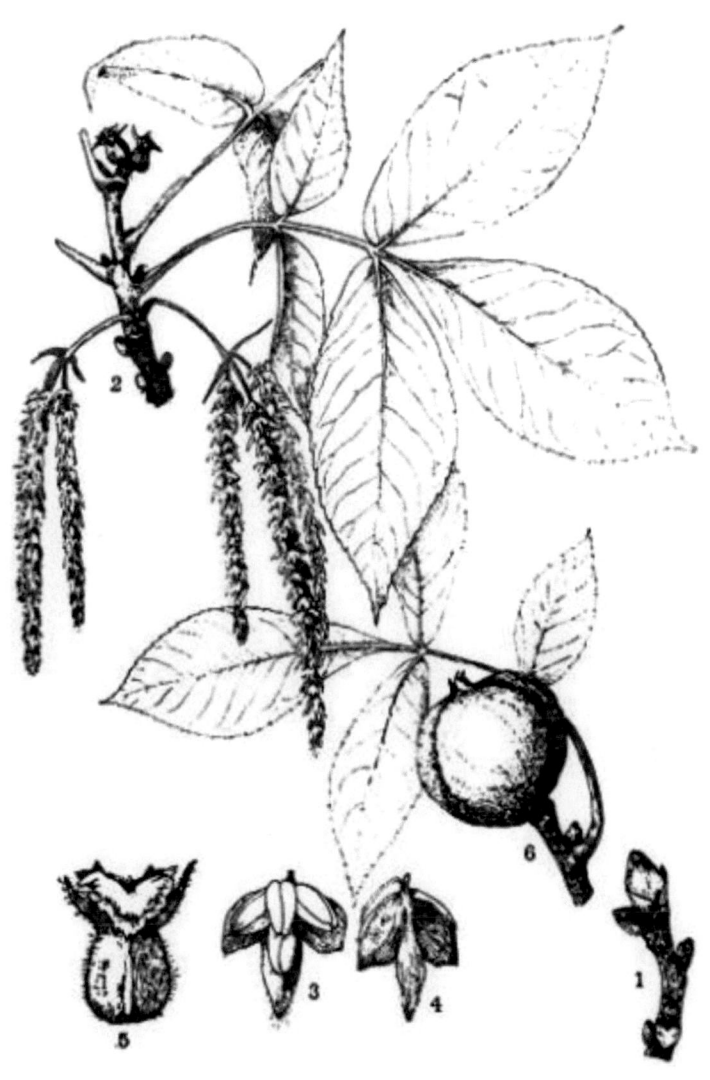

Plate XXIV. — Carya alba.

1. Winter buds.
2. Flowering branch.
3. Sterile flower, front view.
4. Sterile flower, back view.
5. Fertile flower.
6. Fruiting branch.

Carya tomentosa, Nutt.

Hicoria alba, Britton.

Mockernut. White-heart Hickory. Walnut.

Habitat and Range. — In various soils; woods, dry, rocky ridges, mountain slopes.

Niagara peninsula and westward.

Maine and Vermont, — not reported; New Hampshire, — sparingly along the coast; Massachusetts, — rather common eastward; Rhode Island and Connecticut, — common.

South to Florida, ascending 3500 feet in Virginia; west to Kansas, Nebraska, Missouri, Indian territory, and Texas. [Pg 52]

Habit. — A tall and rather slender tree, 50-70 feet high, with a diameter above the swell of the roots of 2-3 feet; attaining much greater dimensions south and west; trunk erect, not shaggy, separating into a few rather large limbs and sending out its upper branches at a sharp angle, forming a handsome, wide-spreading, pyramidal head.

Bark. — Bark of trunk dark gray, thick, hard, close, and rough, becoming narrow-rugged-furrowed; crinkly on small trunks and branches; leaf-scars prominent; season's shoots stout, brown, downy or dusty puberulent, dotted, resinous-scented.

Winter Buds and Leaves. — Buds large, yellowish-brown, ovate, downy. Leaves pinnately compound, alternate, 15-20 inches long; rachis large, downy, swollen at the base; stipules none; leaflets 7-9, opposite, large, yellowish-green and smooth above, beneath paler and thick-downy, at least when young, turning to a clear yellow or russet brown in autumn, the three upper obovate, the two lower ovate, all the leaflets slightly serrate or entire, pointed, base acute to rounded, nearly sessile except the odd one. Aromatic when bruised.

Inflorescence. — May. Sterile and fertile flowers on the same tree, appearing when the leaves are fully grown, — sterile at the base of the season's shoots, in slender, pendulous, downy catkins, 4-8 inches long, usually in threes, branching umbel-like from a common peduncle; scales 3-lobed, hairy; calyx adnate; stamens 4 or 5, anthers red, bearded at the tip: fertile flowers on peduncles at the end of the season's shoots; calyx toothed, hairy, adherent to ovary; corolla none; stigmas 2, hairy.

Fruit. — October. Generally sessile on terminal peduncles, single or in pairs, as large or larger than the fruit of the shagbark, or as small as that of the pignut, oblong-globose to globose: husk hard and thick, separating in 4 segments nearly to the base, strong-scented: nut globular, 4-ridged near the top, thick-shelled: kernel usually small, sweet, edible. The superior size of the fruit and the smallness of the kernel probably give rise to the common name, "mockernut."

Horticultural Value. — Hardy throughout New England; prefers a rich, well-drained soil, but grows well in rocky, [Pg 53] ledgy, exposed situations, and is seldom disfigured by insect enemies. Young trees have large, deep roots, and are difficult to transplant successfully unless they have been frequently transplanted in nurseries, from which, however, they are seldom obtainable. Propagated from seed.

149

Plate XXV. — Carya tomentosa.

1. Winter buds.
2. Flowering branch.
3. Sterile flower, front view.
4. Sterile flower, side view.
5. Sterile flower, top view.
6. Fertile flower, side view.
7. Fruiting branch.

Carya porcina, Nutt.

Hicoria glabra, Britton.

Pignut. White Hickory.

Habitat and Range. — Woods, dry hills, and uplands.

Niagara peninsula and along Lake Erie.

Maine, — frequent in the southern corner of York county; New Hampshire, — common toward the coast and along the lower Merrimac valley; abundant on hills near the Connecticut river, but only occasional above Bellows Falls; Vermont, — Marsh Hill, Ferrisburgh (Brainerd); W. Castleton and Pownal (Eggleston); Massachusetts, — common eastward; along the Connecticut river valley and some of the tributary valleys more common than the shagbark; Rhode Island and Connecticut, — common.

South to the Gulf of Mexico; west to Minnesota, Nebraska, Kansas, Indian territory, and Texas.

Habit. — A stately tree, 50-65 feet high, reaching in the Ohio basin a height of 120 feet; trunk 2-5 feet in diameter, gradually tapering, surmounted by a large, oblong, open, rounded, or pyramidal head, often of great beauty. [Pg 54]

Bark. — Bark of trunk dark ash-gray, uniformly but very coarsely roughened, in old trees smooth or broken into rough and occasionally projecting plates; branches gray; leaf-scars rather prominent; season's shoots smooth or nearly so, purplish changing to gray, with numerous dots.

Winter Buds and Leaves. — Lateral buds smaller than in *C. tomentosa*, oblong, pointed; terminal, globular, with rounded apex; scales numerous, the inner reddish, lengthening to 1 or 2 inches, not dropping till after expansion of the leaves. Leaves pinnately compound, alternate, 10-18 inches long; petiole long and smooth; stipules none; leaflets 5-7, opposite, 2-5 inches long, yellowish-green above, paler beneath, turning to an orange brown in autumn, smooth on both sides; outline, the three upper obovate, the two lower oblong-lanceolate, all taper-pointed; base obtuse, sometimes acute, especially in the odd leaflet.

Inflorescence. — May. Sterile and fertile flowers on the same tree, appearing when the leaves are fully grown, — sterile at the base of the season's shoots, in pendulous, downy, slender catkins, 3-5 inches long, usually in threes, branching umbel-like from a common peduncle; scales 3-lobed, nearly glabrous, lobes of nearly equal length, pointed, the middle narrower; stamens mostly 4, anthers yellowish, beset with white hairs: fertile flowers at the ends of the season's shoots; calyx 4-toothed, pubescent, adherent to the ovary; corolla none; stigmas 2.

Fruit. — October. Single or in pairs, sessile on a short, terminal stalk, shape and size extremely variable, pear-shaped, oblong, round, or obovate, usually about 1½ inches in diameter: husk thin, green turning to brown, when ripe parting in four sections to the center and sometimes nearly to the base: nut rather thick-shelled, not ridged, not sharp-pointed: kernel much inferior in flavor to that of the shagbark.

Horticultural Value. — Hardy throughout New England; grows in all well-drained soils, but prefers a deep, rich loam; a desirable tree for ornamental plantations, especially in lawns, as the deep roots do not interfere with the growth of grass above them; ill-adapted, like all the hickories, for streets, as [Pg 55] the nuts are liable to cause trouble; less readily obtainable in nurseries than the shellbark hickory and equally difficult to transplant. Propagated from the seed.

155

Plate XXVI. — Carya porcina.

1. Winter buds.
2. Flowering branch.
3, 4. Sterile flower, back view.
5. Fertile flower, side view.
6. Fruiting branch.

Carya amara, Nutt.

Hicoria minima, Britton.

Bitternut. Swamp Hickory.

Habitat and Range. — In varying soils and situations; wet woods, low, damp fields, river valleys, along roadsides, occasional upon uplands and hill slopes.

From Montreal west to Georgian bay.

Maine, — southward, rare; New Hampshire, — eastern limit in the Connecticut valley, where it ranges farther north than any other of our hickories, reaching Well's river (Jessup); Vermont, — occasional west of the Green mountains and in the southern Connecticut valley; Massachusetts, — rather common, abundant in the vicinity of Boston; Rhode Island and Connecticut, — common.

South to Florida, ascending 3500 feet in Virginia; west to Minnesota, Nebraska, Kansas, Indian territory, and Texas.

Habit. — A tall, slender tree, 50-75 feet high and 1 foot-2½ feet in diameter at the ground, reaching greater dimensions southward. The trunk, tapering gradually to the point of branching, develops a capacious, spreading head, usually widest near the top, with lively green, finely cut foliage of great beauty, turning to a rich orange in autumn. Easily recognized in winter by its flat, yellowish buds. [Pg 56]

Bark. — Bark of trunk gray, close, smooth, rarely flaking off in thin plates; branches and branchlets smooth; leaf-scars prominent; season's shoots yellow, smooth, yellow-dotted.

Winter Buds and Leaves. — Terminal buds long, yellow, flattish, often scythe-shaped, pointed, with a granulated surface; lateral buds much smaller, often ovate or rounded, pointed. Leaves pinnately compound, alternate, 12-15 inches long; rachis somewhat enlarged at base; stipules none; leaflets 5-11, opposite, 5-6 inches long, 1-2 inches wide, bright green and smooth above, paler and smooth or somewhat downy beneath, turning to orange yellow in autumn; outline lanceolate, or narrowly oval to oblong-obovate, serrate; apex taper-pointed to scarcely acute; base obtuse or rounded except that of the terminal leaflet, which is acute; sessile and inequilateral, except in terminal leaflet, which has a short stem and is equal-sided; sometimes scarcely distinguishable from the leaves of *C. porcina*; often decreasing regularly in size from the upper to the lower pair.

Inflorescence. — May. Sterile and fertile flowers on the same tree, appearing when the leaves are fully grown, — sterile at the base of the season's shoots, or sometimes from the lateral buds of the preceding season, in slender, pendulous catkins, 3-4 inches long, usually in threes, branching umbel-like from a common peduncle; scale 3-lobed, hairy-glandular, middle lobe about the same length as the other two but narrower, considerably longer toward the end of the catkin; stamens mostly 5, anthers bearded at the tip: fertile flowers on peduncles at the end of the season's shoots; calyx 4-lobed, pubescent, adherent to the ovary; corolla none; stigmas 2.

Fruit. — October. Single or in twos or threes at the ends of the branchlets, abundant, usually rather small, about 1 inch long, the width greater than the length; occasionally larger and somewhat pear-shaped: husk separating about to the middle into four segments, with sutures prominently winged at the top or almost to the base, or nearly wingless: nut usually thin-shelled: kernel white, sweetish at first, at length bitter.

Horticultural Value. — Hardy throughout New England; grows almost anywhere, but prefers a rich, loamy or gravelly [Pg 57] soil. A most graceful and attractive hickory, which is transplanted more readily and grows rather more rapidly than the shagbark or pignut, but more inclined than either of these to show dead branches. Seldom for sale by nurserymen or collectors. Grown readily from seed.

Plate XXVII. — Carya amara.

1. Winter bud.
2. Flowering branch.
3. Sterile flower, back view.
4. Sterile flower, front view.
5. Fertile flower.
6. Fruiting branch.

BETULACEÆ. BIRCH FAMILY.

Ostrya Virginica, Willd.

Ostrya Virginiana, Willd.

Hop Hornbeam. Ironwood. Leverwood.

Habitat and Range. — In rather open woods and along highlands.

Nova Scotia to Lake Superior.

Common in all parts of New England.

Scattered throughout the whole country east of the Mississippi, ranging through western Minnesota to Nebraska, Kansas, Indian territory, and Texas.

Habit. — A small tree, 25-40 feet high and 8-12 inches in diameter at the ground, sometimes attaining, without much increase in height, a diameter of 2 feet; trunk usually slender; head irregular, often oblong or loosely and rather broadly conical; lower branches sometimes slightly declining at the extremities, but with branchlets mostly of an upward tendency; spray slender and rather stiff. Suggestive, in its habit, of the elm; in its leaves, of the black birch; and in its fruit, of clusters of hops. [Pg 58]

Bark. — Trunk and large limbs light grayish-brown, very narrowly and longitudinally ridged, the short, thin segments in old trees often loose at the ends; the smaller branches, branchlets, and in late fall the season's shoots, dark reddish-brown.

Winter Buds and Leaves. — Buds small, oblong, pointed, invested with reddish-brown scales. Leaves simple, alternate, roughish, 2-4 inches long, 1-2 inches wide, more or less appressed-pubescent on both sides, dark green above, lighter beneath; outline ovate to oblong-ovate, sharply and for the most part doubly serrate; apex acute to acuminate; base slightly and narrowly heart-shaped, rounded or truncate, mostly with unequal sides; leafstalks short, pubescent; stipules soon falling.

Inflorescence. — April to May. Sterile flowers from wood of the preceding season, lateral or terminal, in drooping, cylindrical catkins, usually in threes; scales broad, laterally rounded, sharp-pointed, ciliate, each subtending several nearly sessile stamens, filaments sometimes forked, with anthers bearded at the tip: fertile catkins about 1 inch in length, on short leafy shoots, spreading; bracts lanceolate, tapering to a long point, ciliate, each subtending two ovaries, each ovary with adherent calyx, enclosed in a hairy bractlet; styles 2, long, linear.

Fruit. — Early September. A small, smooth nut, enclosed in the distended bract; the aggregated fruit resembling a cluster of hops.

Horticultural Value. — Hardy throughout New England; prefers dry or well-drained slopes in gravelly or rocky soil; graceful and attractive, but of rather slow growth; useful in shady situations and worthy of a place in ornamental plantations, but too small for street use. Seldom raised by nurserymen; collected plants moved with difficulty. Propagated from seed.

167

Plate XXVIII. — Ostrya Virginica.

1. Winter buds.
2. Flowering branch.
3. Sterile flower, back view.
4. Sterile flower, front view.
5. Fertile catkin.
6. Fertile flower.
7. Fruiting branch.

[Pg 59]

Carpinus Caroliniana, Walt.

Hornbeam. Blue Beech. Ironwood. Water Beech.

Habitat and Range. — Low, wet woods, and margins of swamps.

Province of Quebec to Georgian bay.

Rather common throughout New England, less frequent towards the coast.

South to Florida; west to Minnesota, Nebraska, Kansas, Indian territory, and Texas.

Habit. — A low, spreading tree, 10-30 feet high, with a trunk diameter of 6-12 inches, rarely reaching 2 feet; trunk short, often given a fluted appearance by projecting ridges running down from the lower branches to the ground; in color and smoothness resembling the beech; lower branches often much declined, upper going out at various angles, often zigzag but keeping the same general direction; head wide, close, flat-topped to rounded, with fine, slender spray.

Bark. — Trunk smooth, close, dark bluish-gray; branchlets grayish; season's shoots light green turning brown, more or less hairy.

Winter Buds and Leaves. — Leaf-buds small, oval or ovoid, acute to obtuse. Leaves simple, alternate, 2-3 inches long, dull green above, lighter beneath, turning to scarlet or crimson in autumn;

outline ovate or slightly obovate oblong or broadly oval, irregularly and sharply doubly serrate; veins prominent and pubescent beneath, at least when young; apex acuminate to acute; base rounded, truncate, acute, or slightly and unevenly heart-shaped; leafstalk rather short, slender, hairy; stipules pubescent, falling early.

Inflorescence.—May. Sterile flowers from growth of the preceding season in short, stunted-looking, lateral catkins, mostly single; scales ovate or rounded, obtuse, each subtending several stamens; filaments very short, mostly 2-forked; anthers bearded at the tip: fertile flowers at the ends of leafy shoots of the season, in loose catkins; bractlets foliaceous, [Pg 60] each subtending a green, ovate, acute, ciliate, deciduous scale, each scale subtending two pistils with long reddish styles.

Fruit.—In terminal catkins made conspicuous by the pale green, much enlarged, and leaf-like 3-lobed bracts, each bract subtending a dark-colored, sessile, striate nutlet.

Horticultural Value.—Hardy throughout New England; prefers moist, rich soil, near running water, on the edges of wet land or on rocky slopes in shade. Its irregular outline and curiously ridged trunk make it an interesting object in landscape plantations. It is not often used, however, because it is seldom grown in nurseries, and collected plants do not bear removal well. Propagated from the seed.

171

Plate XXIX. — Carpinus Caroliniana.

1. Winter buds.
2. Flowering branch.
3. Sterile flower, back view.
4. Sterile flower, front view.
5. Fertile catkin.
6. Fertile flower.
7. Fruiting branch.

BETULA.

Inflorescence. — In scaly catkins, sterile and fertile on the same tree, appearing with or before the leaves from shoots of the previous season, — sterile catkins terminal and lateral, formed in summer, erect or inclined in the bud, drooping when expanded in the following spring; sterile flowers usually 3, subtended by a shield-shaped bract with 2 bractlets; each flower consisting of a 1-scaled calyx and 2 anthers, which appear to be 4 from the division of the filaments into two parts, each of which bears an anther cell: fertile catkins erect or inclined at the end of very short leafy branchlets; fertile flowers subtended by a 3-lobed bract falling with the nuts; bractlets none; calyx none; corolla none; consisting of 2-3 ovaries crowned with 2 spreading styles. [Pg 61]

Betula lenta, L.

Black Birch. Cherry Birch. Sweet Birch.

Habitat and Range. — Moist grounds; rich woods, old pastures, fertile hill-slopes, banks of rivers.

Newfoundland and Nova Scotia to the Lake Superior region.

Maine, — frequent; New Hampshire, — in the highlands of the southern section, and along the Connecticut river valley to a short distance north of Windsor; Vermont, — frequent in the western part of the state, and in the southern Connecticut valley (*Flora of Vermont*, 1900); Massachusetts and Rhode Island,—frequent throughout, especially in the highlands, less often near the coast; Connecticut,—widely distributed, especially in the Connecticut river valley, but not common.

South to Delaware, along the mountains to Florida; west to Minnesota and Kansas.

Habit.—A medium-sized or rather large tree, 50-75 feet high, with a trunk diameter of 1-4 feet, often conspicuous along precipitous ledges, springing out of crevices in the rocks and assuming a variety of picturesque forms. In open ground the dark trunk develops a symmetrical, wide-spreading, hemispherical head broadest at its base, the lower limbs horizontal or drooping sometimes nearly to the ground. The limbs are long and slender, often more or less tortuous, and separated ultimately into a delicate, polished spray. Distinguished by its long purplish-yellow, pendulous catkins in spring, and in summer by its glossy, bright green, and abundant foliage, which becomes yellow in autumn.

Bark.—Bark of trunk on old trees very dark, separating and cleaving off in large, thickish plates; on young trees and on branches a dark reddish brown, not separating into thin layers, smooth, with numerous horizontal lines 1-3 inches long; branchlets reddish-brown, shining, with shorter lateral lines; season's shoots with small, pale dots. Inner bark very aromatic, having a strong checkerberry flavor,—hence the common [Pg 62] name, "checkerberry birch"; called also "cherry birch," from the resemblance of its bark to that of the garden cherry.

Winter Buds and Leaves.—Buds reddish-brown, oblong or conical, pointed, inner scales whitish, elongating as the bud opens. Leaves simple, in alternate pairs, 3-4 inches long and one-half as wide, shining green above and downy when young, paler beneath and silvery-downy along the prominent, straight veins; outline ovate-oval, ovate-oblong, or oval; sharply serrate to doubly serrate; apex acute to acuminate; base heart-shaped to obtuse; leafstalk short, often curved, hairy when young; stipules soon falling.

Inflorescence.—April to May. Sterile catkins 3-4 inches long, slender, purplish-yellow; scales fringed: fertile catkins erect or suberect, sessile or nearly so, ½-1 inch long, oblong-cylindrical; bracts pubescent; lateral lobes wider than in *B. lutea.*

Fruit.—Fruiting catkins oblong-cylindrical, nearly erect; bracts with 3 short, nearly equal diverging lobes: nut obovate-oblong, wider than its wings; upper part of seed-body usually appressed-pubescent.

Horticultural Value.—Hardy throughout New England; grows everywhere from swamps to hilltops, but prefers moist rocky slopes and a loamy or gravelly soil; occasionally offered by nurserymen; both nursery and collected plants are moved without serious difficulty; apt to grow rather unevenly.

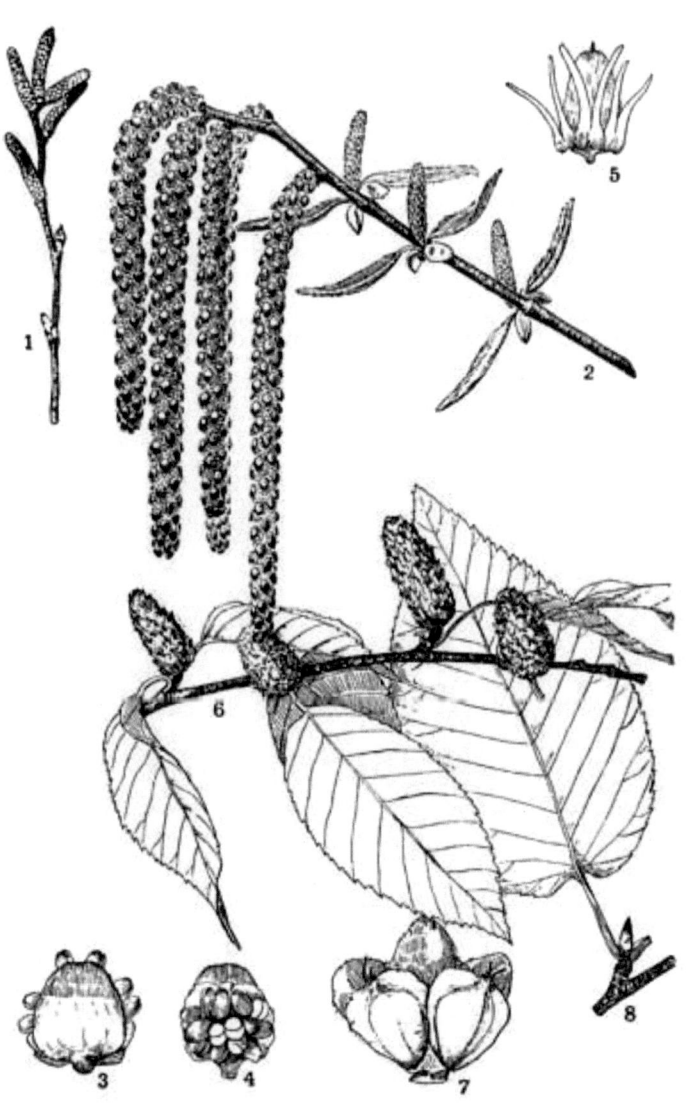

Plate XXX. — Betula lenta.

1. Winter buds.
2. Flowering branch.
3. Sterile flower, back view.
4. Sterile flower, front view.
5. Fertile flower.
6. Fruiting branch.
7. Fruit.
8. Mature leaf.

[Pg 63]

Betula lutea, Michx. f.

Yellow Birch. Gray Birch.

Habitat and Range. — Low, rich woodlands, mountain slopes.

Newfoundland and Nova Scotia to Rainy river.

New England, — abundant northward; common throughout, from borders of lowland swamps to 1000 feet above the sea level; more common at considerable altitudes, where it often occurs in extensive patches or belts.

South to the middle states, and along the mountains to Tennessee and North Carolina; west to Minnesota.

Habit. — A large tree, at its maximum in northern New England 60-90 feet high and 2-4 feet in diameter at the base. In the forest the main trunk separates at a considerable height into a few large branches which rise at a sharp angle, curving slightly, forming a rather small, irregular head, widest near the top; while in open ground the head is broad-spreading, hemispherical, with numerous rather equal, long and slender branches, and a fine spray with drooping tendencies. In the sunlight the silvery-yellow feathering and the metallic sheen of trunk and branches make the yellow birch one of the most attractive trees of the New England forest.

Bark. — Bark of trunks and large limbs in old trees gray or black-ish, lustreless, deep-seamed, split into thick plates, standing out at all sorts of angles; in trees 6-8 inches in diameter, scarf-bark lustrous, parted in ribbon-like strips, detached at one end and running up the trunk in delicate, tattered fringes; season's shoots light yellowish-green, minutely buff-dotted, woolly-pubescent, becoming in successive seasons darker and more lustrous, the dots elongating into horizontal lines. Aromatic but less so than the bark of the black birch; not readily detachable like the bark of the canoe birch.

Winter Buds and Leaves. — Buds conical, ¼ inch long, mostly appressed, tips of scales brownish. Leaves simple, in alternate pairs or scattered singly along the stem; 3-5 inches [Pg 64] long, ½-2 inches wide, dull green on both sides, paler beneath and more or less pubescent on the straight veins; outline oval to oblong, for the most part doubly serrate; apex acuminate or acute; base heart-shaped, obtuse or truncate; leafstalk short, grooved, often pubescent or woolly; stipules soon falling.

Inflorescence. — April to May. Sterile catkins 3-4 inches long, purplish-yellow; scales fringed: fertile catkins sessile or nearly so, about 1 inch long, cylindrical; bracts 3-lobed, nearly to the middle, pubescent, lobes slightly spreading.

Fruit. — Fruiting catkins oblong or oblong-ovoid, about 1 inch long and two-thirds as thick, erect: nut oval to narrowly obovate, tapering at each end, pubescent on the upper part, about the width of its wing.

Horticultural Value. — Hardy throughout New England; grows in wet or dry situations, but prefers wet, peaty soil, where its roots can find a constant supply of moisture; similar to the black birch, equally valuable in landscape-gardening, but less desirable as a street tree; transplanted without serious difficulty.

Differences between black birch and yellow birch:

Black Birch. — Bark reddish-brown, not separable into thin layers; leaves bright green above, finely serrate; fruiting catkins cylindrical; bark of twigs decidedly aromatic.

Yellow Birch. — Bark yellow, separable into thin layers; leaves dull green above; serration coarser and more decidedly doubly

serrate; fruiting catkins ovoid or oblong-ovoid; flavor of bark less
distinctly aromatic.

Plate XXXI. — Betula lutea.

1. Winter buds.
2. Flower-buds.
3. Flowering branch.
4-6. Sterile flowers.
7. Fertile flower.
8. Bract.
9. Fruiting branch.
10. Fruit.

[Pg 65]

Betula nigra, L.

Red Birch. River Birch.

Habitat and Range. — Along rivers, ponds, and woodlands inundated a part of the year.

Doubtfully and indefinitely reported from Canada.

No stations in Maine, Vermont, Rhode Island, or Connecticut; New Hampshire, — found sparingly along streams in the southern part of the state; abundant along the banks of Beaver brook, Pelham (F. W. Batchelder); Massachusetts, — along the Merrimac river and its tributaries, bordering swamps in Methuen and ponds in North Andover.

South, east of the Alleghany mountains, to Florida; west, locally through the northern tier of states to Minnesota and along the Gulf states to Texas; western limits, Nebraska, Kansas, Indian territory, and Missouri.

Habit. — A medium-sized tree, 30-50 feet high, with a diameter at the ground of 1-1½ feet; reaching much greater dimensions southward. The trunk, frequently beset with small, leafy, reflexed branchlets, and often only less frayed and tattered than that of the yellow birch, develops a light and feathery head of variable outline, with numerous slender branches, the upper long and drooping, the reddish spray clothed with abundant dark-green foliage.

Bark. — Reddish, more or less separable into layers, fraying into shreddy, cinnamon-colored fringes; in old trees thick, dark reddish-brown, and deeply furrowed; branches dark red or cinnamon, giving rise to the name of "red birch"; season's shoots downy, pale-dotted.

Winter Buds and Leaves. — Buds small, mostly appressed near the ends of the shoots, tapering at both ends. Leaves simple, alternate, 3-4 inches long, two-thirds as wide, dark green and smooth above, paler and soft-downy beneath, turning bright yellow in autumn; outline rhombic-ovate, with unequal and sharp double serratures; leafstalk short and downy; stipules soon falling. [Pg 66]

Inflorescence. — April to May. Sterile catkins usually in threes, 2-4 inches long, scales 2-3-flowered: fertile catkins bright green, cylindrical, stalked; bracts 3-lobed, the central lobe much the longest, tomentose, ciliate.

Fruit. — June. Earliest of the birches to ripen its seed; fruiting catkins 1-2 inches long, cylindrical, erect or spreading; bracts with the 3 lobes nearly equal in width, spreading, the central lobe the longest: nut ovate to obovate, ciliate.

Horticultural Value. — Hardy throughout New England; grows in all soils, but prefers a station near running water; young trees grow vigorously and become attractive objects in landscape plantations; especially useful along river banks to bind the soil; retains its lower branches better than the black or yellow birches. Seldom found in nurseries, and rather hard to transplant; collected plants do fairly well.

Plate XXXII. — Betula nigra.

1. Leaf-buds.
2. Flower-buds.
3. Branch with sterile and fertile catkins.
4. Sterile flower.
5. Fertile flower.
6. Scale of fertile flower.
7. Fruit.
8. Fruiting branch.

Betula populifolia, Marsh.

White Birch. Gray Birch. Oldfield Birch. Poplar Birch. Poverty Birch. Small White Birch.

Habitat and Range. — Dry, gravelly soils, occasional in swamps and frequent along their borders, often springing up on burnt lands.

Nova Scotia to Lake Ontario.

Maine, — abundant; New Hampshire, — abundant eastward, as far north as Conway, and along the Connecticut to Westmoreland; Vermont, — common in the western and frequent in the southern sections; Massachusetts, Rhode Island, and Connecticut, — common. [Pg 67]

South, mostly in the coast region, to Delaware; west to Lake Ontario.

Habit. — A small tree, 20-35 feet high, with a diameter at the ground of 4-8 inches, occasionally much exceeding these dimensions; under favorable conditions, of extreme elegance. The slender, seldom erect trunk, continuous to the top of the tree, throws out numerous short, unequal branches, which form by repeated subdivisions a profuse, slender spray, disposed irregularly in tufts or masses, branches and branchlets often hanging vertically or drooping at the ends. Conspicuous in winter by the airy lightness of the narrow open head and by the contrast of the white trunk with the dark spray; in summer, when the sun shines and the air stirs, by the delicacy, tremulous movement, and brilliancy of the foliage.

Bark. — Trunk grayish-white, with triangular, dusty patches below the insertion of the branches; not easily separable into layers; branches dark brown or blackish; season's shoots brown, with numerous small round dots becoming horizontal lines and increasing in length with the age of the tree. The white of the bark does not readily come off upon clothing.

Winter Buds and Leaves. — Buds somewhat diverging from the twig; narrow conical or cylindrical, reddish-brown. Leaves simple, alternate, single or in pairs, 3-4 inches long, two-thirds as wide, bright green above, paler beneath, smooth and shining on both sides, turning to a pale shining yellow in autumn, resinous, glandular-dotted when young; outline triangular, coarsely and irregularly doubly serrate; apex taper-pointed; base truncate, heart-shaped, or acute; leafstalks long and slender; stipules dropping early.

Inflorescence. — May. Sterile catkins usually solitary or in pairs, slender-cylindrical, 2-3 inches long: fertile catkins erect, green, stalked; bracts minutely pubescent.

Fruit. — Fruiting catkins erect or spreading, cylindrical, about 1¼ inches long and ½ inch in diameter, stalked; scales 3-parted above the center, side lobes larger, at right angles or reflexed: nuts small, ovate to obovate, narrower than the wings, combined wings from broadly obcordate to butterfly-shape, wider than long. [Pg 68]

Horticultural Value. — Hardy throughout New England, growing in every kind of soil, finest specimens in deep, rich loam. Were this tree not so common, its graceful habit and attractive bark would be more appreciated for landscape gardening; only occasionally grown by nurserymen, best secured through collectors; young collected plants, if properly selected, will nearly all live.

187

Plate XXXIII. — Betula populifolia.

1. Branch with sterile and fertile catkins.
2. Sterile flower, back view.
3. Fertile flower.
4. Scale of fertile flower.
5. Fruiting branch.
6. Fruit.

Betula papyrifera, Marsh.

Canoe Birch. White Birch. Paper Birch.

Habitat and Range. — Deep, rich woods, river banks, mountain slopes.

Canada, Atlantic to Pacific, northward to Labrador and Alaska, to the limit of deciduous trees.

Maine, — abundant; New Hampshire, — in all sections, most common on highlands up to the alpine area of the White mountains, above the range of the yellow birch; Vermont, — common; Massachusetts, — common in the western and central sections, rare towards the coast; Rhode Island, — not reported; Connecticut, — occasional in the southern sections, frequent northward.

South to Pennsylvania and Illinois; west to the Rocky mountains and Washington on the Pacific coast.

Var. *minor*, Tuckerman, is a dwarf form found upon the higher mountain summits of northern New England.

Habit. — A large tree, 50-75 feet high, with a diameter of 1-3 feet; occasionally of greater dimensions. The trunk develops a broad-spreading, open head, composed of a few large limbs ascending at an acute angle, with nearly horizontal [Pg 69] secondary branches and a slender, flexible spray without any marked tendency to droop. Characterized by the dark metallic lustre of the branchlets, the dark green foliage, deep yellow in autumn, and the chalky whiteness of the trunk and large branches; a singularly picturesque tree, whether standing alone or grouped in forests.

Bark.—Easily detachable in broad sheets and separable into thin, delicately colored, paper-like layers, impenetrable by water, outlasting the wood it covers. Bark of trunk and large branches chalky-white when fully exposed to the sun, lustreless, smooth or ragged-frayed, in very old forest trees encrusted with huge lichens, and splitting into broad plates; young trunks and smaller branches smooth, reddish or grayish brown, with numerous roundish buff dots which enlarge from year to year into more and more conspicuous horizontal lines. The white of the bark readily rubs off upon clothing.

Winter Buds and Leaves.—Buds small, ovate, flattish, acute to rounded. Leaves simple, alternate, 3-5 inches long, two-thirds as wide, dark green and smooth above, beneath pale, hairy along the veins, sometimes in young trees thickly glandular-dotted on both sides; outline ovate, ovate-oblong, or ovate-orbicular, more or less doubly serrate; apex acute to acuminate; base somewhat heart-shaped, truncate or obtuse; leafstalk 1-2 inches long, grooved above, downy; stipules falling early.

Inflorescence.—April to May. Sterile catkins mostly in threes, 3-4 inches long: fertile catkins 1-1½ inches long, cylindrical, slender-peduncled, erect or spreading; bracts puberulent.

Fruit.—Fruiting catkins 1-2 inches long, cylindrical, short-stalked, spreading or drooping: nut obovate to oval, narrower than its wings; combined wings butterfly-shaped, nearly twice as wide as long.

Horticultural Value.—Hardy throughout New England; prefers a well-drained loam or gravelly soil, but does fairly well in almost any situation; young trees rapid growing and vigorous, but with the same tendency to grow irregularly that [Pg 70] is shown by the black and yellow birches; transplanted without serious difficulty; not offered by many nurserymen, but may be obtained from northern collectors.

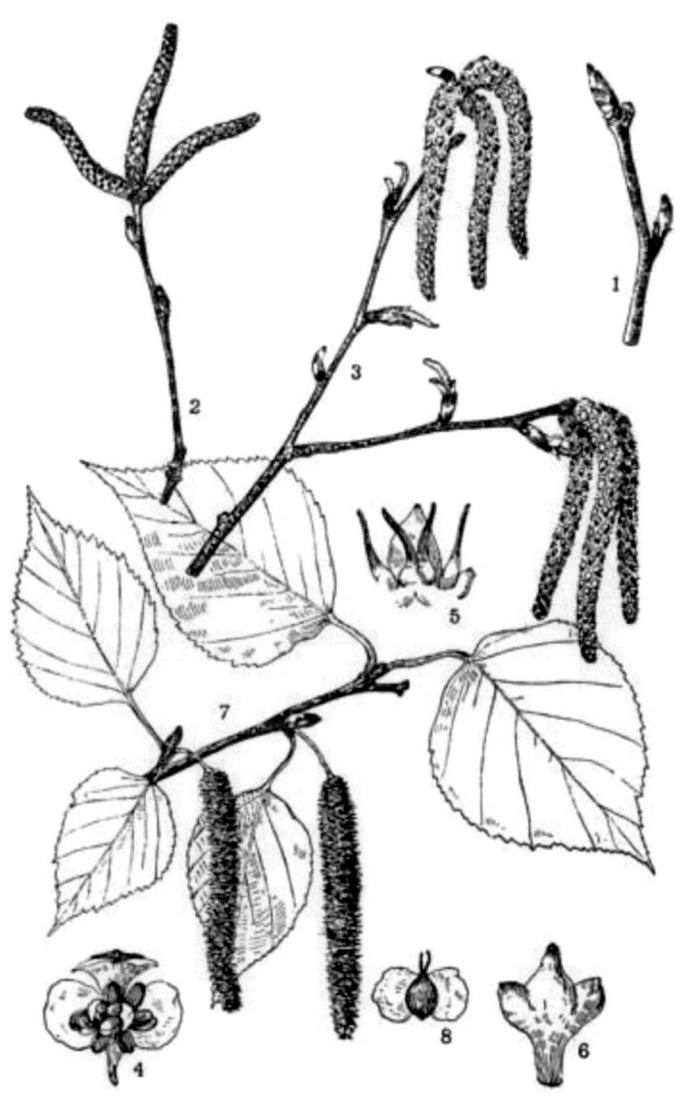

Plate XXXIV. — Betula papyrifera.

1. Leaf-buds.
2. Flower-buds.
3. Flowering branch.
4. Sterile flower, front view.
5. Fertile flower, front view.
6. Scale of fertile flower.
7. Fruiting branch.
8. Fruit.

Alnus glutinosa, Medic.

European Alder.

This is the common alder of Great Britain and central Europe southward, growing chiefly along water courses, in boggy grounds and upon moist mountain slopes; introduced into the United States and occasionally escaping from cultivation; sometimes thoroughly established locally. In Medford, Mass., there are many of these plants growing about two small ponds and upon the neighboring lowlands, most of them small, but among them are several trees 30-40 feet in height and 8-12 inches in diameter at the ground, distinguishable at a glance from the shrubby native alders by their greater size, more erect habit, and darker trunks.

FAGACEÆ. BEECH FAMILY.

Fagus ferruginea, Ait.

Fagus Americana, Sweet. Fagus atropunicea, Sudw.

Beech.

Habitat and Range. — Moist, rocky soil.

Nova Scotia through Quebec and Ontario.

Maine, — abundant; New Hampshire, — throughout the state; common on the Connecticut-Merrimac watershed, enters largely into the composition of the hardwood forests of Coos county; Vermont, — abundant; Massachusetts, [Pg 71] — in western sections abundant, common eastward; Rhode Island and Connecticut, — common.

South to Florida; west to Wisconsin, Missouri, and Texas.

Habit. — A tree of great beauty, rising to a height of 50-75 feet, with a diameter at the ground of 1½-4 feet; under favorable conditions attaining much greater dimensions; trunk remarkably smooth, sometimes fluted, in the forests tall and straight, in open situations short and stout; head symmetrical, of various shapes, — rounded, oblong, or even obovate; branches numerous, mostly long and slender, curving slightly upward at their tips, near the point of branching horizontal or slightly drooping, beset with short branchlets which form a flat, dense, and beautiful spray; roots numerous, light brown, long, and running near the surface. Tree easily distinguishable in winter by the dried brownish-white leaves, spear-like buds, and smooth bark.

Bark. — Trunk light blue gray, smooth, unbroken, slightly corrugated in old trees, often beautifully mottled in blotches or bands and invested by lichens; branches gray; branchlets dark brown and smooth; spray shining, reddish-brown; season's shoots a shining olive green, orange-dotted.

Winter Buds and Leaves. — Buds conspicuous, long, very slender, tapering slowly to a sharp point; scales rich brown, lengthening as the bud opens. Leaves set in plane of the spray, simple, alternate, 3-5 inches long, one-half as wide, silky-pubescent with fringed edges when young, nearly smooth when fully grown, green on both sides, turning to rusty yellows and browns in autumn, persistent till midwinter; outline oval, serrate; apex acuminate; base rounded; veins strong, straight, terminating in the teeth; leafstalk short, hairy at first; stipules slender, silky, soon falling.

Inflorescence. — May. Appearing with the leaves from the season's shoots, sterile flowers from the lower axils, in heads suspended at the end of silky threads 1-2 inches long; calyx campanulate, pubescent, yellowish-green, mostly 6-lobed; petals none; stamens 6-16; anthers exserted; ovary wanting or abortive: fertile flowers from the upper axils, usually single or [Pg 72] in pairs, at the end of a short peduncle; involucre 4-lobed, fringed with prickly scales; calyx with six awl-shaped lobes; ovary 3-celled; styles 3.

Fruit. — A prickly bur, thick, 4-valved, splitting nearly to the base when ripe: nut sharply triangular, sweet, edible.

Horticultural Value. — Hardy throughout New England; grows well in any good soil, but prefers deep, rich, well-drained loam; usually obtainable in nurseries; when frequently transplanted, safely moved. Its clean trunk and limbs, deep shade, and freedom from insect pests make it one of the most attractive of our large trees for use, summer or winter, in landscape gardening; few plants, however, will grow beneath it; the bark is easily disfigured; it has a bad habit of throwing out suckers and is liable to be killed by any injury to the roots. Propagated from the seed. The purple beech, weeping beech, and fern-leaf beech are well-known horticultural forms.

Plate XXXV. — Fagus ferruginea.

1. Winter buds.
2. Flowering branch.
3. Sterile flower.
4. Fertile flower.
5. Fruiting branch.
6. Section of fruit.
7. Nut.

Castanea sativa, var. Americana, Watson and Coulter.

Castanea dentata, Borkh. Castanea vesca, var. Americana, Michx.

Chestnut.

Habitat and Range. — In strong, well-drained soil; pastures, rocky woods, and hillsides.

Ontario, — common.

Maine, — southern sections, probably not indigenous north of latitude 44° 20'; New Hampshire, — Connecticut valley near the river, as far north as Windsor, Vt.; most abundant in the Merrimac valley south of Concord, but occasional a [Pg 73] short distance northward; Vermont, — common in the southern sections, especially in the Connecticut valley; occasional as far north as Windsor (Windsor county), West Rutland (Rutland county), Burlington (Chittenden county); Massachusetts, — rather common throughout the state, but less frequent near the sea; Rhode Island and Connecticut, — common.

South to Delaware, along the mountains to Alabama; west to Michigan, Indiana, and Tennessee.

Habit. — A tree of the first magnitude, rising to a height of 60-80 feet and reaching a diameter of 5-6 feet above the swell of the roots, with a spread sometimes equaling or even exceeding the height; attaining often much greater proportions. The massive trunk separates usually a few feet from the ground into several stout horizontal or ascending branches, the limbs higher up, horizontal or rising at a broad angle, forming a stately, open, roundish, or inversely pyramidal head; branchlets slender; spray coarse and not abundant; foliage bright green, dense, casting a deep shade; flowers profuse, the long, sterile catkins upon their darker background of leaves conspicuous upon the hill slopes at a great distance. A tree that may well dispute precedence with the white or red oak.

Bark. — Bark of trunk in old trees deeply cleft with wide ridges, hard, rough, dark gray; in young trees very smooth, often shining; season's shoots green or purplish-brown, white-dotted.

Winter Buds and Leaves. — Buds small, ovate, brown, acutish. Leaves simple, alternate, 5-10 inches long, 1-3 inches wide, bright clear green above, paler beneath and smooth on both sides; outline oblong-lanceolate, sharply and coarsely serrate; veins straight, terminating in the teeth; apex acuminate; base acute or obtuse; leafstalk short; stipules soon falling.

Inflorescence. — June to July. Appearing from the axils of the season's shoots, after the leaves have grown to their full size; sterile catkins numerous, clustered or single, erect or [Pg 74] spreading, 4-10 inches long, slender, flowers pale yellowish-green or creamcolored; calyx pubescent, mostly 6-parted; stamens 15-20; odor offensive when the anthers are discharging their pollen: fertile flowers near the base of the upper sterile catkins or in separate axils, 1-3 in a prickly involucre; calyx 6-toothed; ovary ovate, styles as many as the cells of the ovary, exserted.

Fruit. — Burs round, thick, prickly, 2-4 inches in diameter, opening by 4 valves: nuts 1-5, dark brown, covered with whitish down at apex, flat on one side when there are several in a cluster, ovate when only one, sweet and edible.

Horticultural Value. — Hardy throughout New England; prefers fertile, well-drained, gravelly or rocky soil; rather difficult to transplant; usually obtainable in nurseries. Its vigorous and rapid growth, massive, broad-spreading head and attractive flowers make it a valuable tree for landscape gardening, but in public places the prickly burs and edible fruit are a serious disadvantage. Propagated from the seed.

203

Plate XXXVI.—Castanea sativa, var. Americana.

1. Winter buds.
2. Flowering branch.
3. Sterile flower.
4. Fertile flower.
5. Fruit.
6. Nut.

QUERCUS.

Inflorescence appearing with the leaves in spring; sterile catkins from terminal or lateral buds on shoots of the preceding year, bracted, usually several in a cluster, unbranched, long, cylindrical, pendulous; bracts of sterile flowers minute, soon falling; calyx parted or lobed; stamens 3-12, undivided: fertile flowers terminal or axillary upon the new shoots, single or few-clustered, bracted, erect; involucre scaly, becoming the cupule or cup around the lower part of the acorn; ovary 3-celled; stigma 3-lobed. [Pg 75]

White Oaks.

Leaves with obtuse or rounded lobes or teeth; cup-scales thickened or knobbed at base; stigmas sessile or nearly so; fruit maturing the first year.

Black Oaks.

Leaves with pointed or bristle-tipped lobes and teeth; cup-scales flat; stigmas on spreading styles; fruit maturing the second year.

Quercus alba, L.

White Oak.

Habitat and Range. — Light loams, sandy plains, and gravelly ridges, often constituting extensive tracts of forest.

Quebec and Ontario.

Maine, — southern sections; New Hampshire, — most abundant eastward; in the Connecticut valley confined to the hills in the immediate vicinity of the river, extending up the tributary streams a short distance and disappearing entirely before reaching the mouth of the Passumpsic (W. F. Flint); Vermont, — common west of the Green mountains, less so in the southern Connecticut valley (*Flora of Vermont*, 1900); Massachusetts, Rhode Island, and Connecticut, — common.

South to the Gulf of Mexico; west to Minnesota, Nebraska, Kansas, Arkansas, and Texas.

Habit. — A tree of the first rank, 50-75 feet high and 1-6 feet in diameter above the swell of the roots, exhibiting considerable diversity in general appearance, trunk sometimes dissolving into branches like the American elm, and sometimes continuous to the top. The finest specimens in open land are characterized by a rather short, massive trunk, with stout, horizontal, far-reaching limbs, conspicuously gnarled and twisted in old age, forming a wide-spreading, [Pg 76] open head of striking grandeur, the diameter at the base of which is sometimes two or three times the height of the tree.

Bark. — Trunk and larger branches light ash-gray, sometimes nearly white, broken into long, thin, loose, irregular, soft-looking flakes; in old trees with broad, flat ridges; inner bark light; branchlets ash-gray, mottled; young shoots grayish-green, roughened with minute rounded, raised dots.

Winter Buds and Leaves.—Buds ⅛ to ¼ inch long, round-ovate, reddish-brown. Leaves simple, alternate, 3-7 inches long, 2-4 inches wide, delicately reddish-tinted and pubescent upon both sides when young; at maturity glabrous, light dull or glossy green above, paler and somewhat glaucous beneath, turning to various reds in autumn; outline obovate to oval; lobes 5-9; ascending, varying greatly in different trees; when few, short and wide-based, with comparatively shallow sinuses; when more in number, ovate-oblong, with deeper sinuses, or somewhat linear-oblong, with sinuses reaching nearly to midrib; apex of lobe rounded; base of leaf tapering; leafstalks short; stipules linear, soon falling. The leaves of this species are often persistent till spring, especially in young trees.

Inflorescence.—May. Appearing when the leaves are half grown; sterile catkins 2-3 inches long, with slender, usually pubescent thread; calyx yellow, pubescent; lobes 5-9, pointed: pistillate flowers sessile or short-peduncled, reddish, ovate-scaled.

Fruit.—Maturing in the autumn of the first year, single, or more frequently in pairs, sessile or peduncled: cup hemispherical to deep saucer shaped, rather thin; scales rough-knobby at base. acorn varying from ½ inch to an inch in length, oblong-ovoid: meat sweet and edible, said to be when boiled a good substitute for chestnuts.

Horticultural Value.—Hardy in New England; grows well in all except very wet soils, in all open exposures and in light shade; like all oaks, difficult to transplant unless prepared by frequent transplanting in nurseries, from which it is not readily obtainable in quantity; grows very slowly and nearly [Pg 77] uniformly up to maturity; comparatively free from insect enemies but occasionally disfigured by fungous disease which attacks immature leaves in spring. Propagated from seed.

Plate XXXVII.—Quercus alba.

1. Winter buds.
2. Flowering branch.
3-4. Sterile flower, front view.
5. Fertile flower, side view.
6. Fruiting branch.
7-8. Variant leaves.

Quercus stellata, Wang.

Q. obtusiloba, Michx. Q. minor, Sarg.

Post Oak. Box White Oak.

Habitat and Range.

Doubtfully reported from southern Ontario.

In New England, mostly in sterile soil near the sea-coast; Massachusetts, — southern Cape Cod from Falmouth to Brewster, the most northern station reported, occasional; the islands of Naushon, Martha's Vineyard where it is rather common, and Nantucket where it is rare; Rhode Island, — along the shore of the northern arm of Wickford harbor (L. W. Russell); Connecticut, — occasional along the shores of Long Island sound west of New Haven.

South to Florida; west to Kansas, Indian territory, and Texas.

Habit. — Farther south, a tree of the first magnitude, reaching a height of 100 feet, with a trunk diameter of 4 feet; in southern New England occasionally attaining in woodlands a height of 50-60 feet; at its northern limit in Massachusetts, usually 10 to 35 feet in height, with a diameter at the ground of 6-12 inches. The trunk throws out stout, tough, and often conspicuously crooked branches, the lower horizontal or declining, forming a disproportionately large head, with dark green, dense foliage. Near the shore the limbs often grow very low, stretching along the ground as if from an underground stem. [Pg 78]

Bark. — Resembling that of the white oak, but rather a darker gray, rougher and firmer; upon old trunks furrowed and cut into oblongs; small limbs brownish-gray, rough-dotted; season's shoots densely tawny-tomentose.

Winter Buds and Leaves. — Buds small, rounded or conical, brownish, scales minutely pubescent or scurfy. Leaves simple, alternate, 3-8 inches long, two-thirds as wide, thickish, yellowish-green and tomentose upon both sides when young, becoming a deep, somewhat glossy green above, lighter beneath, both sides still somewhat scurfy; general outline of leaf and of lobes, and number and shape of the latter, extremely variable; type-form 5-lobed, all the lobes rounded, the three upper lobes much larger, more or less subdivided, often squarish, the two lower tapering to an acute, rounded, or truncate base; sinuses deep, variable, often at right angles to the midrib; leafstalk short, tomentose; stipules linear, pubescent, occasionally persistent till midsummer. The leaves are often arranged at the tips of the branches in star-shaped clusters, giving rise to the specific name *stellata*.

Inflorescence. — May. Sterile catkins 1-3 inches long, connecting thread woolly; calyx 4-8 parted, lobes acute, densely pubescent, yellow; stamens 4-8, *anthers with scattered hairs*: pistillate flowers single or in clusters of 2, 3, or more, sessile or on a short stem; stigma red.

Fruit. — Maturing the first season, single and sessile, or nearly so, or in clusters of 2, 3, or more, on short footstalks: cup top-shaped or cup-shaped, ⅓-½ the length of the acorn, about ¾ inch wide, thin; scales smooth or sometimes hairy along the top, acutish or roundish, slightly thickened at base: acorn ½-1 inch long, sweet.

Horticultural Value. — Hardy in New England; prefers a good, well-drained, open soil; quite as slow-growing as the white oak; seldom found in nurseries and difficult to transplant. Propagated from the seed.

217

Plate XXXVIII.—Quercus stellata.

1. Winter buds.
2. Flowering branch.
3. Sterile flower, back view.
4. Sterile flower, front view.
5. Fertile flower.
6. Fruiting branch.

[Pg 79]

Quercus macrocarpa, Michx.

Bur Oak. Over-cup Oak. Mossy-cup Oak.

Habitat and Range.—Deep, rich soil; river valleys.

Nova Scotia to Manitoba, not attaining in this region the size of the white oak, nor covering as large areas.

Maine,—known only in the valleys of the middle Penobscot (Orono) and the Kennebec (Winslow, Waterville); Vermont,—lowlands about Lake Champlain, especially in Addison county, not common; Massachusetts,—valley of the Ware river (Worcester county), Stockbridge and towns south along the Housatonic river (Berkshire county); Rhode Island,—no station reported; Connecticut,—probably introduced in central and eastern sections, possibly native near the northern border.

South to Pennsylvania and Tennessee; west to Montana, Nebraska, Kansas, Indian territory, and Texas.

Habit.—A medium-sized tree, 40-60 feet high, with a trunk diameter of 1-3 feet; attaining great size in the Ohio and Mississippi river basins; trunk erect, branches often changing direction, ascending, save the lowest, which are often nearly horizontal; branchlets numerous, on the lowest branches often declined or drooping; head wide-spreading, rounded near the center, very rough in aspect;

distinguished in summer by the luxuriance of the dark-green foliage and in autumn by the size of its acorns.

Bark.—Bark of trunk and branches ash-gray, but darker than that of the white oak, separating on old trees into rather firm, longitudinal ridges; bark of branches sometimes developed into conspicuous corky, wing-like layers; season's shoots yellowish-brown, minutely hairy, with numerous small, roundish, raised dots.

Winter Buds and Leaves.—Buds brown, 1/16 to ⅛ inch long, conical, scattered along the shoots and clustered at the enlarged tips. Leaves simple, alternate, 6-9 inches long, 3-4 inches broad, smooth and dark green above, lighter and [Pg 80] downy beneath; outline obovate to oblong, varying from irregularly and deeply sinuate-lobed, especially near the center, to nearly entire, base wedge-shaped; stalk short; stipules linear, pubescent.

Inflorescence.—May. Sterile catkins 3-5 inches long; calyx mostly 5-parted, yellowish-green; divisions linear-oblong, more or less persistent; stamens 10; anthers yellow, glabrous: pistillate flowers sessile or short-stemmed; scales reddish; stigma red.

Fruit.—Maturing the first season; extremely variable; sessile or short-stemmed: cup top-shaped to hemispherical, 3/4-2 inches in diameter, with thick, close, pointed scales, the upper row often terminating in a profuse or sparing hairy or leafy fringe: acorn ovoid, often very large, sometimes sunk deeply and occasionally entirely in the cup.

Horticultural Value.—Hardy in New England; in general appearance resembling the swamp white oak, but better adapted to upland; grows rather slowly in any good, well-drained soil; difficult to transplant; seldom disfigured by insects or disease; occasionally grown in nurseries. Propagated from seed. A narrower-leafed form with small acorns (var. *olivæformis*) is occasionally offered.

Plate XXXIX. — Quercus macrocarpa.

1. Winter buds.
2. Flowering branch.
3. Sterile flower, back view.
4. Sterile flower, front view.
5. Fertile flowers.
6. Fruiting branch.

Quercus bicolor, Willd.

Quercus platanoides, Sudw.

Swamp White Oak.

Habitat and Range.—In deep, rich soil; low, moist, fertile grounds, bordering swamps and along streams.

Quebec to Ontario, where it is known as the blue oak.

Maine,—York county; New Hampshire,—Merrimac valley as far as the mouth of the Souhegan, and probably through [Pg 81] out Rockingham county; Vermont,—low grounds about Lake Champlain; Massachusetts,—frequent in the western and central sections, common eastward; Rhode Island and Connecticut,—common.

South to Delaware and along the mountains to northern Georgia; west to Minnesota, Iowa, east Kansas, and Arkansas.

Habit.—A medium-sized tree, 40-60 feet high, with a trunk diameter of 2-3 feet; attaining southward of the Great Lakes and in the Ohio basin much greater dimensions; roughest of all the oaks, except the bur oak, in general aspect; trunk erect, continuous, in young trees often beset at point of branching with down-growing, scraggly branchlets, surmounted by a rather regular pyramidal head, the lower branches horizontal or declining, often descending to the ground, with a short, stiff, abundant, and bushy spray; smaller twigs ridgy, widening beneath buds; foliage a dark shining green; heads of large trees less regular, rather open, with a general resemblance to the head of the white oak, but narrower at the base, with less contorted limbs.

Bark.—Bark of trunk and larger branches thick, dark grayish-brown, longitudinally striate, with flaky scales; bark of young stems, branches, and branchlets darker, separating in loose scales which curl back, giving the tree its shaggy aspect; season's shoots yellowish-green.

Winter Buds and Leaves.—Buds brown, roundish-ovate, obtuse. Leaves simple, alternate, 3-8 inches long, 2-4 wide, downy on both sides when unfolding, at maturity thick and firm, smooth and dark shining green above, slightly to conspicuously whitish-downy beneath, in autumn brownish-yellow; obovate, coarsely and deeply crenate or obtusely shallow-lobed, when opening sometimes pointed and tapering to a wedge-shaped base, often constricted near the center; leafstalk short; stipules linear, soon falling.

Inflorescence. — May. Sterile catkins 2-3 inches long, thread hairy; calyx deeply 3-7-parted, pale yellow, hairy; stamens 5-8; anthers yellow, glabrous: pistillate flowers tomentose, on rather long, hairy peduncles; stigmas red. [Pg 82]

Fruit. — Variable, on stems 1-3 inches long, maturing the first season, single or frequently in twos: cup rounded, rather thin, deep, rough to mossy, often with fringed margins: acorn about 1 inch long, oblong-ovoid, more or less tapering: meat sweet, edible.

Horticultural Value. — Hardy throughout New England; grows in any good soil, wet or dry, but prefers a position on the edge of moist or boggy land, where its roots can find a constant supply of water; growth fairly rapid; seldom affected by insects or disease; occasionally offered by nurserymen and rather less difficult to transplant than most of the oaks. Its sturdy, rugged habit and rich dark green foliage make it a valuable tree for ornamental plantations or even for streets.

Plate XL. — Quercus bicolor.

1. Winter buds.
2. Flowering branch.
3. Sterile flower, side view.
4. Sterile flower, front view.
5. Fertile flowers.
6. Fruiting branch.

Quercus Prinus, L.

Chestnut Oak. Rock Chestnut Oak.

Habitat and Range. — Woods, rocky banks, hill slopes.

Along the Canadian shore of Lake Erie.

Maine, — Saco river and Mt. Agamenticus, near the southern coast (York county); New Hampshire, — belts or patches in the eastern part of the state and along the southern border, Hinsdale, Winchester, Brookline, Manchester, Hudson; Vermont, — western part of the state throughout, not common; abundant at Smoke mountain at an altitude of 1300 feet, and along the western flank of the Green mountains, at least in Addison county; Massachusetts, — eastern sections, Sterling, Lancaster, Russell, Middleboro, rare in Medford and Sudbury, frequent on the Blue hills; Rhode Island, — locally common; Connecticut, — common. [Pg 83]

South to Delaware and along the mountains to Georgia, extending nearly to the summit of Mt. Pisgah in North Carolina; west to Kentucky, Tennessee, and Alabama.

Habit. — A small or medium-sized tree, 25-50 feet high, with a trunk diameter of 1-2½ feet, assuming noble proportions southward, often reaching a height of 75-100 feet and trunk diameter of 5-6 feet; trunk tall, straight, continuous to the top of the tree, scarcely tapering to the point of ramification, surmounted by a spacious, open head.

Bark. — Bark of trunk and large branches deep gray to dark brown or blackish, in firm, broad, continuous ridges, with small, close surface scales; bark of young trees and of branchlets smooth, brown, and more or less lustrous; season's shoots light brown.

Winter Buds and Leaves. — Buds ovate to cylindrical, mostly acute, brownish. Leaves simple, alternate, 5-8 inches long, 2-5 inches wide, dark green and smooth above, paler and more or less downy beneath; outline obovate to oval, undulate-crenate; apex blunt-pointed; base wedge-shaped, obtuse or slightly rounded, often unequal-sided; veins straight, parallel, prominent beneath; leafstalk ½-1½ inches long; stipules linear, soon falling.

Inflorescence. — May. Sterile catkins 2-3 inches long; calyx 5-9-parted, yellow, hairy; divisions oblong, densely pubescent; stamens 5-9; anthers yellow, glabrous: pistillate flowers with hairy scales and dark red stigmas.

Fruit. — Seldom abundant, maturing the first season, variable in size, on stems usually equal to or shorter than the leaf-stems: cup thin, hemispheric or somewhat top-shaped, deep; scales small, knobby-thickened at the base: acorns ¾-1½ inches long, ovoid-conical, sweet.

Horticultural Value. — Hardy throughout New England; prefers a light gravelly or stony soil; rapid-growing and free from disease; more easily and safely transplanted than most oaks; occasionally offered by nurserymen, who propagate it from the seed. Its vigorous, clean habit of growth and handsome foliage should give it a place in landscape gardening and street use. [Pg 84]

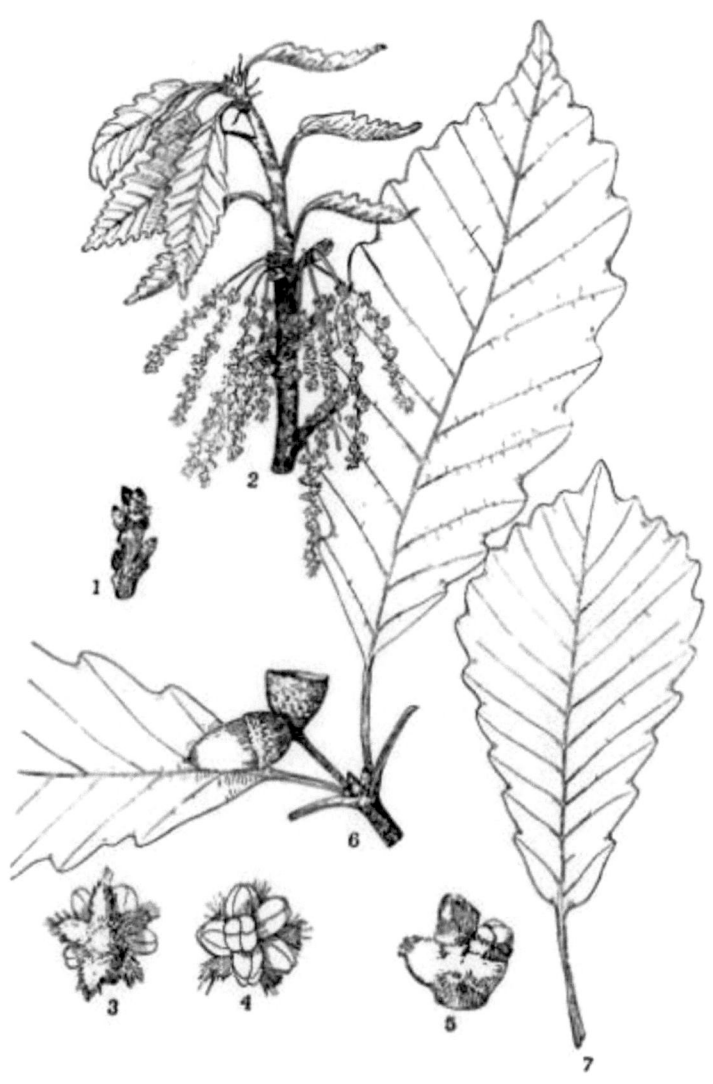

231

Plate XLI. — Quercus Prinus.

1. Winter buds.
2. Flowering branch.
3. Sterile flower, back view.
4. Sterile flower, front view.
5. Fertile flowers.
6. Fruiting branch.
7. Variant leaf.

Quercus Muhlenbergii, Engelm.

Quercus acuminata, Sarg.

Chestnut Oak.

Habitat and Range. — Dry hillsides, limestone ridges, rich bottoms.

Ontario.

Vermont, — Gardner's island, Lake Champlain; Ferrisburg (Pringle); Connecticut, — frequent (J. N. Bishop, 1895); on the limestone formation in the neighborhood of Kent (Litchfield county, C. K. Averill); often confounded by collectors with *Q. Prinus*; probably there are other stations. Not authoritatively reported from the other New England states.

South to Delaware and District of Columbia, along the mountains to northern Alabama; west to Minnesota, Nebraska, Kansas, Indian territory, and Texas.

Habit. — A medium-sized tree, 30-40 feet high, with a trunk diameter of 1-2 feet, attaining much greater dimensions in the basins of the Ohio, Mississippi, and their tributaries; trunk in old trees enlarged at the base, erect, branches rather short for the genus, forming a narrow oblong or roundish head.

Bark. — Bark of trunk and large branches grayish or pale ash-colored, comparatively thin, flaky; branchlets grayish-brown; season's shoots in early summer purplish-green with pale dots.

Winter Buds and Leaves. — Buds ovate, acute to obtuse, brownish. Leaves simple, alternate; in the typical form as recognized by Muhlenburg, 3-6 inches long, 1½-2 inches wide, [Pg 85] glossy dark green above, pale and minutely downy beneath; outline lanceolate or lanceolate-oblong, with rather equal, coarse, sharp, and often inflexed teeth; apex acuminate; base wedge-shaped or acute; stipules soon falling. There is also a form of the species in which the leaves are much larger, 5-7 inches in length and 3-5 inches in width, broadly ovate or obovate, with rounded teeth; distinguishable from *Q. Prinus* only by the bark and fruit.

Inflorescence. — May. Appearing with the leaves; sterile catkins 2-4 inches long; calyx yellow, hairy, segments 5-8, ciliate; stamens 5-8, anthers yellow: pistillate flowers sessile or on short spikes; stigma red.

Fruit. — Maturing the first season, sessile or short-peduncled: cup covering about half the nut, thin, shallow, with small, rarely much thickened scales: acorn ovoid or globose, about 3/4 inch long.

Horticultural Value. — Hardy in New England; grows in all good dry or moist soils, in open or partly shaded situations; maintains a nearly uniform rate of growth till maturity, and is not seriously affected by insects. It forms a fine individual tree and is useful in forest plantations. Propagated from seed.

237

Plate XLII.—Quercus Muhlenbergii.

1. Winter buds.
2. Flowering branch.
3. Sterile flower.
4. Fertile flowers.
5. Fruiting branch.

Quercus prinoides, Willd.

Scrub White Oak. Scrub Chestnut Oak.

More or less common throughout the states east of the Mississippi; westward apparently grading into *Q. Muhlenbergii*, within the limits of New England mostly a low shrub, rarely assuming a tree-like habit. The leaves vary from rather narrow-elliptical to broadly obovate, are rather regularly and coarsely toothed, bright green and often lustrous on the upper surface. [Pg 86]

Quercus rubra, L.

Red Oak.

Habitat and Range.—Growing impartially in a great variety of soils, but not on wet lands.

Nova Scotia and New Brunswick to divide west of Lake Superior.

Maine,—common, at least south of the central portions; New Hampshire,—extending into Coos county, far north of the White mountains; Vermont, Massachusetts, Rhode Island, and Connecticut,—common; probably in most parts of New England the most common of the genus; found higher up the slopes of mountains than the white oak.

South to Tennessee, Virginia, and along mountain ranges to Georgia; reported from Florida; west to Minnesota, Nebraska, Kansas, and Texas.

Habit.—The largest of the New England oaks, 50-85 feet high, with a diameter of 2-6 feet above the swell of the roots; occasionally attaining greater dimensions; trunk usually continuous to the top of the tree, often heavily buttressed; point of branching higher than in the white oak; branches large, less contorted, and rising at a sharper angle, the lower sometimes horizontal; branchlets rather slender; head extremely variable, in old trees with ample space for growth, open, well-proportioned, and imposing; sometimes oblong in outline, wider near the top, and sometimes symmetrically rounded, not so broad, however, as the head of the white oak; conspicuous in summer by its bright green, abundant foliage, which turns to dull purplish-red in autumn.

Bark.—Bark of trunk and lower parts of branches in old trees dark gray, firmly, coarsely, and rather regularly ridged, smooth elsewhere; in young trees greenish mottled gray, smooth throughout; season's shoots at first green, taking a reddish tinge in autumn, marked with pale, scattered dots.

Winter Buds and Leaves. Buds ovate, conical, sharp pointed. Leaves simple, alternate, 4-8 inches long, 3-5 inches broad, bright green above, paler beneath, dull brown [Pg 87] in autumn; outline oval or obovate, sometimes scarcely distinguishable by the character of its lobing from *Q. tinctoria*; in the typical form, lobes broadly triangular or oblong, with parallel sides bristle-pointed; leafstalks short; stipules linear, soon falling.

Inflorescence.—Earliest of the oaks, appearing in late April or early May, when the leaves are half-grown; sterile catkins 3-5 inches long; calyx mostly 4-lobed; lobes rounded; stamens mostly 4; anthers yellow: pistillate flowers short-stemmed; calyx lobes mostly 3 or 4; stigmas long, spreading.

Fruit.—Maturing in the second year, single or in pairs, sessile or short-stalked: cup sometimes turbinate, usually saucer-shaped with a flat or rounded base, often contracted at the opening and surmounted by a kind of border; scales closely imbricated, reddish-brown, more or less downy, somewhat glossy, triangular-acute to obtuse, pubescent: acorn nearly cylindrical or ovoid, tapering to a broad, rounded top.

Horticultural Value. — Hardy throughout New England; grows in all well-drained soils, but prefers a rich, moist loam; more readily obtainable than most of our oaks; in common with other trees of the genus, nursery trees must be transplanted frequently to be moved with safety; grows rapidly and is fairly free from disfiguring insects; the oak-pruner occasionally lops off its twigs. When once established, it grows as rapidly as the sugar maple, and is worthy of much more extended use in street and landscape plantations. Propagated from the seed.

Plate XLIII.—Quercus rubra.

1. Winter buds.
2. Flowering branch.
3. Sterile flower.
4. Fertile flowers, side view.
5. Fruiting branch.

[Pg 88]

Quercus coccinea, Wang.

Scarlet Oak.

Habitat and Range.—Most common in dry soil.

Ontario.

Maine,—valley of the Androscoggin, southward; New Hampshire and Vermont,—not authoritatively reported by recent observers; Massachusetts,—more common in the eastern than western sections, sometimes covering considerable areas; Rhode Island and Connecticut,—common.

South to the middle states and along the mountains to North Carolina and Tennessee; reported from Florida; west to Minnesota, Nebraska, and Missouri.

Habit.—A medium-sized tree, 30-50 feet high and 1-3 feet in trunk diameter; attaining greater dimensions southward; trunk straight and tapering, branches regular, long, comparatively slender, not contorted, the lower nearly horizontal, often declined at the ends; branchlets slender; head open, narrow-oblong or rounded, graceful; foliage deeply cut, shining green in summer and flaming scarlet in autumn; the most brilliant and most elegant of the New England oaks.

Bark.—Trunk in old trees dark gray, roughly and firmly ridged; inner bark red; young trees and branches smoothish, often marked with dull red seams and more or less mottled with gray.

Winter Buds and Leaves. — Buds small, reddish-brown, ovate to oval, acutish, partially hidden by enlarged base of petiole. Leaves simple, alternate, extremely variable, more commonly 3-6 inches long, two-thirds as wide, bright green and shining above, paler beneath, smooth on both sides but often with a tufted pubescence on the axils beneath, turning scarlet in autumn, deeply lobed, the rounded sinuses sometimes reaching nearly to the midrib; lobes 5-9, rather slender and set at varying angles, sparingly toothed and bristly tipped; apex acute; base truncate to acute; leafstalk 1-1½ inches long, slender, swollen at base. [Pg 89]

Inflorescence. — Early in May. Appearing when the leaves are half grown; sterile catkins 2-4 inches long; calyx most commonly 4-parted; pubescent; stamens commonly 4, exserted; anthers yellow, glabrous: pistillate flowers red; stigmas long, spreading, reflexed.

Fruit. — Maturing in the autumn of the second year, single or in twos or threes, sessile or on rather short footstalks: cup top-shaped or cup-shaped, about half the length of the acorn, occasionally nearly enclosing it, smooth, more or less polished, thin-edged; scales closely appressed, firm, elongated, triangular, sides sometimes rounded, homogeneous in the same plant: acorn ½-¾ inch long, variable in shape, oftenest oval to oblong: kernel white within; less bitter than kernel of the black oak.

Horticultural Value. — Hardy throughout New England; grows in any light, well-drained soil, but prefers a fertile loam. Occasionally offered by nurserymen, but as it is disposed to make unsymmetrical young trees it is not grown in quantity, and it is not desirable for streets. Its rapid growth, hardiness, beauty of summer foliage, and its brilliant colors in autumn make it desirable in ornamental plantations. Propagated from the seed.

Plate XLIV. — Quercus coccinea.

1. Winter buds.
2. Flowering branch.
3. Sterile flowers, side view.
4. Fertile flower, side view.
5. Fruiting branch.

Quercus velutina, Lam.

Quercus tinctoria, Bartram. Quercus coccinea, **var.** *tinctoria, Gray.*

Black Oak. Yellow Oak.

Habitat and Range.—Poor soils; dry or gravelly uplands; rocky ridges.

Southern and western Ontario.

Maine,—York county; New Hampshire,—valley of the lower Merrimac and eastward, absent on the highlands, [Pg 90] reappearing within three or four miles of the Connecticut, ceasing at North Charlestown; Vermont,—western and southeastern sections; Massachusetts,—abundant eastward; Rhode Island and Connecticut,—frequent.

South to the Gulf states; west to Minnesota, Kansas, Indian territory, and Texas.

Habit.—One of our largest oaks, 50-75 feet high and 2-4 feet in diameter, exceptionally much larger, attaining its maximum in the Ohio and Mississippi basins; resembling *Q. coccinea* in the general disposition of its mostly stouter branches; head wide-spreading, rounded; trunk short; foliage deep shining green, turning yellowish or reddish brown in autumn.

Bark.—Bark of trunk dark gray or blackish, often lighter near the seashore, thick, usually rough near the ground even in young trees, in old trees deeply furrowed, separating into narrow, thick, and firmly adherent block-like strips; inner bark thick, yellow, and bitter; branches and branchlets a nearly uniform, mottled gray; season's shoots scurfy-pubescent.

Winter Buds and Leaves.—Buds ⅛-¼ inch long, bluntish to pointed, conspicuously clustered at ends of branches. Leaves simple, alternate, of two forms so distinct as to suggest different species, *a* (Plate XLV, 8) varying towards *b* (Plate XLV, 6), and *b* often scarcely distinguishable from the leaf of the scarlet oak; in both forms outline obovate to oval, lobes usually 7, densely woolly when opening, more or less pubescent or scurfy till midsummer or later, dark shining green above, lighter beneath, becoming brown or dull red in autumn.

Form *a*, sinuses shallow, lobes broad, rounded, mucronate.

Form *b*, sinuses deep, extending halfway to the midrib or farther, oblong or triangular, bristle-tipped.

Inflorescence.—Early in May. Appearing when the leaves are half grown; sterile catkins 2-5 inches long, with slender, pubescent threads; calyx usually 3-4-lobed; lobes ovate, acute to rounded, hairy-pubescent; stamens 3-7, commonly 4-5; anthers yellow: pistillate flowers reddish, pubescent, at first nearly sessile; stigmas 3, red, divergent, reflexed. [Pg 91]

Fruit.—Maturing the second year; nearly sessile or on short foot-stalks: cup top-shaped to hemispherical; scales less firm than in *Q. coccinea*, tips papery and transversely rugulose, obtuse or rounded, or some of them acutish, often lacerate-edged, loose towards the thick and open edge of the cup: acorn small: kernel yellow within and bitter.

Horticultural Value.—Hardy throughout New England; grows in well-drained soils, but prefers a rich, moist loam; of vigorous and rapid growth when young, but as it soon begins to show dead branches and becomes unsightly, it is not a desirable tree to plant, and is rarely offered by nurserymen. Propagated from seed.

Note.—Apparently runs into *Q. coccinea*, from which it may be distinguished by its rougher and darker trunk, the yellow color and bitter taste of the inner bark, its somewhat larger and more pointed buds, the greater pubescence of its inflorescence, young shoots and leaves, the longer continuance of scurf or pubescence upon the leaves, the yellow or dull red shades of the autumn foliage, and by the yellow color and bitter taste of the nut.

251

Plate XLV. — Quercus velutina.

1. Winter buds.
2. Flowering branch.
3. Sterile flower, 4-lobed calyx.
4. Sterile flower, 3-lobed calyx.
5. Fertile flower.
6. Fruiting branch.
7. Fruit.
8. Variant leaf.

Quercus palustris, Du Roi.

Pin Oak. Swamp Oak. Water Oak.

Habitat and Range. — Low grounds, borders of forests, wet woods, river banks, islets in swamps.

Ontario.

Northern New England, — no station reported; Massachusetts, — Amherst (Stone, *Bull. Torrey Club,* IX, 57; J. E. Humphrey, *Amherst Trees*); Springfield, south to Connecticut, rare; Rhode Island, — southern portions, bordering the great Kingston swamp, and on the margin of the Pawcatuck [Pg 92] river (L. W. Russell); Connecticut, — common along the sound, frequent northward, extending along the valley of the Connecticut river to the Massachusetts line.

South to the valley of the lower Potomac in Virginia; west to Minnesota, east Kansas, Missouri, Arkansas, and Indian territory.

Habit. — A medium-sized tree, 40-50 feet high, with trunk diameter of 1-2 feet, occasionally reaching a height of 60-70 feet (L. W. Russell), but attaining its maximum of 100 feet in height and upward in the basins of the Ohio and Mississippi rivers; trunk rather slender, often fringed with short, drooping branchlets, lower tier of branches short and mostly descending, the upper long, slender, and often beset with short, lateral shoots, which give rise to the common name; head graceful, open, rounded and symmetrical when young, in old age becoming more or less irregular; foliage delicate; bright shining green in autumn, often turning to a brilliant scarlet.

Bark. — Bark of trunk dark, furrowed and broken in old trees, in young trees grayish-brown, smoothish; branchlets shining, light brown.

Winter Buds and Leaves. — Buds short, conical, acute. Leaves simple, alternate, 3-5 inches long, bright green, smooth and shining above, duller beneath, with tufted hairs in the angles of the veins; outline broadly obovate to ovate; lobes divergent, triangular, toothed or entire, bristle-pointed; sinuses broad, rounded; leafstalk slender; stipules linear, soon falling.

Inflorescence. — May. Appearing when the leaves are half grown; sterile catkins 2-4 inches long; segments of calyx mostly 4 or 5, obtuse or rounded, somewhat lacerate; stamens mostly 4 or 5, anthers yellow, glabrous: pistillate flowers with broadly ovate scales; stigmas stout, red, reflexed.

Fruit. — Abundant, maturing the second season, short-stemmed: cup saucer-shaped, with firm, appressed scales, shallow: acorns ovoid to globose, about ½ inch long, often striate, breadth sometimes equal to entire length of fruit.

Horticultural Value. — Probably hardy throughout New England; grows in wet soils, but prefers a rich, moist loam; [Pg 93] of rapid and uniform growth, readily and safely transplanted, and but little disfigured by insects; obtainable in leading nurseries. Propagated from the seed.

Plate XLVI. — Quercus palustris.

1. Winter buds.
2. Flowering branch.
3. Sterile flower, side view.
4. Fertile flower, side view.
5. Fruiting branch.

Quercus ilicifolia, Wang.

Quercus nana, Sarg. Quercus pumila, Sudw.

Scrub Oak. Bear Oak.

Habitat and Range.—In poor soils; sandy plains, gravelly or rocky hills.

Maine,—frequent in eastern and southern sections and upon Mount Desert island; New Hampshire,—as far north as Conway, more common near the lower Connecticut; Vermont,—in the eastern and southern sections as far north as Bellows Falls; Massachusetts, Rhode Island, and Connecticut,—too abundant, forming in favorable situations dense thickets, sometimes covering several acres.

South to Ohio and the mountain regions of North Carolina and Kentucky; west to the Alleghany mountains.

Habit.—Shrub or small tree, usually 3-8 feet high, but frequently reaching a height of 15-25 feet; trunk short, sometimes in peaty swamps 10-13 inches in diameter near the ground, branches much contorted, throwing out numerous branchlets of similar habit, forming a stiff, flattish head; beautiful for a brief week in spring by the delicate greens and reds of the opening leaves and reds and yellows of the numerous catkins. Sometimes associated with *Q. prinoides*.

Bark.—Old trunks dark gray, with small, closely appressed scales; small trunks and branches grayish-brown, not furrowed or scaly; younger branches marked with pale yellow, [Pg 94] raised dots; season's shoots yellowish-green, with a tawny, scurfy pubescence.

Winter Buds and Leaves.—Buds ⅛-¼ inch long, ovoid or conical, covered with imbricated, brownish, minutely ciliate scales. Leaves simple, alternate, 3-4 inches long and 2-3 inches broad; when unfolding reddish above and woolly on both sides, when mature yellowish-green and somewhat glossy above, smooth except on the midrib, rusty-white, and pubescent beneath; very variable in outline and in the number (3-7) and shape of lobes, sometimes entire, oftenest obovate with 5 bristle-tipped angular lobes, the two lower much smaller; base unequal, wedge-shaped, tip obtuse or rounded; leafstalk short; stipules linear, soon falling.

Inflorescence. — Early in May. Appearing when the leaves are half grown; sterile catkins 2-4 inches long; calyx pubescent, lobes oftenest 2-3, rounded; stamens 3-5; anthers red or yellow: pistillate flowers numerous; calyx lobes ovate, pointed, reddish, pubescent; stigmas 3, reddish, recurved, spreading.

Fruit. — Abundant, maturing in the autumn of the second year, clustered along the branchlets on stout, short stems: cup top-shaped or hemispherical: acorn about ½ inch long, varying greatly in shape, mostly ovoid or spherical, brown, often striped lengthwise.

Horticultural Value. — Hardy in New England; grows well in dry, gravelly, ledgy, or sandy soil, where few other trees thrive; useful in such situations where a low growth is required; but as it is not procurable in quantity from nurseries, it must be grown from the seed. The leaves are at times stripped off by caterpillars, but otherwise it is not seriously affected by insects or fungous diseases.

1

4

5

3

2

Plate XLVII. — Quercus ilicifolia.

1. Flowering branch.
2. Sterile flower, side view.
3. Fertile flowers, side view.
4. Fruiting branch.
5. Variant leaves.

[Pg 95]

ULMACEÆ. ELM FAMILY.

Ulmus Americana, L.

Elm. American Elm. White Elm.

Habitat and Range. — Low, moist ground; thrives especially on rich intervales.

From Cape Breton to Saskatchewan, as far north as 54° 30'.

Maine, — common, most abundant in central and southern portions; New Hampshire, — common from the southern base of the White mountains to the sea; in the remaining New England states very common, attaining its highest development in the rich alluvium of the Connecticut river valley.

South to Florida; west to Dakota, Nebraska, Kansas, and Texas.

Habit. — In the fullness of its vigor the American elm is the most stately and graceful of the New England trees, 50-110 feet high and 1-8 feet in diameter above the swell of the roots; characterized by an erect, more or less feathered or naked trunk, which loses itself completely in the branches, by arching limbs, drooping branchlets set at a wide angle, and by a spreading head widest near the top. Modifications of these elements give rise to various well-marked forms which have received popular names.

1. In the vase-shaped tree, which is usually regarded as the type, the trunk separates into several large branches which rise, slowly

diverging, 40-50 feet, and then sweep outward in wide arches, the smaller branches and spray becoming pendent.

2. In the umbrella form the trunk remains entire nearly to the top of the tree, when the branches spread out abruptly, forming a broad, shallow arch, fringed at the circumference with long, drooping branchlets.

3. The slender trunk of the plume elm rises, usually undivided, a considerable height, begins to curve midway, and is capped with a one-sided tuft of branches and delicate, elongated branchlets.

4. The drooping elm differs from the type in the height of the arch and greater droop of the branches, which sometimes sweep the ground. [Pg 96]

5. In the oak form the limbs are more or less tortuous and less arching, forming a wide-spreading, rounded head.

In all forms short, irregular, pendent branchlets are occasional along the trunks. The trees most noticeably feathered are usually of medium size, and have few large branches, the superfluous vitality manifesting itself in a copious fringe, which sometimes invests and obliterates the great pillars which support the masses of foliage. Conspicuous at all seasons of the year,—in spring when its brown buds are swollen to bursting, or when the myriads of flowers, insignificant singly, give in the sunlight an atmosphere of purplish-brown; when clothed with light, airy masses of deep green in summer or pale yellow in autumn, or in winter when the great trunk and mighty sweep of the arching branches distinguish it from all other trees. The roots lie near the surface and run a great distance.

Bark.—Dark gray, irregularly and broadly striate, rather firmly ridged, in very old trees sometimes partially detached in plates; branches ash-gray, smooth; branchlets reddish-brown; season's shoots often pubescent, light brown in late fall.

Winter Buds and Leaves.—Buds ovate, brown, flattened, obtuse to acute, smooth. Leaves simple, alternate, 2-5 inches long, 2-3 inches wide, dark green and roughish above, lighter and downy at first beneath; outline ovate or oval to obovate-oblong, sharply and usually doubly serrate; apex abruptly pointed; base half acute, half rounded, produced on one side, often slightly heart-shaped or obtuse; veins straight and prominent; leafstalk stout, short; stipules small, soon falling. Leaves drop in early autumn.

Inflorescence. — April. In loose lateral clusters along the preceding season's shoots; flowers brown or purplish, mostly perfect, with occasional sterile and fertile on the same tree; stems slender; calyx 7-9-lobed, hairy or smooth; stamens 7-9, filaments slender, anthers exserted, brownish-red; ovary flat, green, ciliate; styles 2.

Fruit. — Ripening in May, before the leaves are fully grown, a samara, ½ inch in diameter, oval or ovate, smooth on both [Pg 97] sides, hairy on the edge, the notch in the margin closed or partially closed by the two incurved points.

Horticultural Value. — Hardy throughout New England; grows in any soil, but prefers a deep, rich loam; the ideal street tree with its high, overarching branches and moderate shade; grows rapidly, throws out few low branches, bears pruning well; now so seriously affected by numerous insect enemies that it is not planted as freely as heretofore; objectionable on the borders of gardens or mowing land, as the roots run along near the surface for a great distance. Very largely grown in nurseries, usually from seed, sometimes from small collected plants. Though so extremely variable in outline, there are no important horticultural forms in cultivation.

266

Plate XLVIII. — Ulmus Americana.

1. Winter buds.
2. Flowering branch.
3. Flower, side view.
4. Fruiting branch.
5. Mature leaf.

Ulmus fulva, Michx.

Ulmus pubescens, Walt.

Slippery Elm. Red Elm.

Habitat and Range.—Rich, low grounds, low, rocky woods and hillsides.

Valley of the St. Lawrence, apparently not abundant.

Maine,—District of Maine (Michaux, *Sylva of North America*, ed. 1853, III, 53), rare; Waterborough (York county, Chamberlain, 1898); New Hampshire,—valley of the Connecticut, usually disappearing within ten miles of the river; ranges as far north as the mouth of the Passumpsic; Vermont,—frequent; Massachusetts,—rare in the eastern sections, frequent westward; Rhode Island.—infrequent; Connecticut,—occasional.

South to Florida; west to North Dakota and Texas.

Habit.—A small or medium-sized tree, 40-60 feet high, with a trunk diameter of 1-2½ feet; head in proportion to the height [Pg 98] of the tree, the widest spreading of the species, characterized by its dark, hairy buds and rusty-green, dense and rough foliage.

Bark.—Bark of trunk brown and in old trees deeply furrowed; larger branches grayish-brown, somewhat striate; branchlets grayish-brown, rough, marked with numerous dots, downy; season's shoots light gray and very rough; inner bark mucilaginous, hence the name "slippery elm."

Winter Buds and Leaves.—Buds ovate to rounded-cylindrical, acute or obtuse, very dark, densely tomentose, very conspicuous just before unfolding. Leaves simple, alternate, 4-8 inches long, 3-4 inches wide, thickish, minutely hairy above and woolly beneath when young, at maturity pale rusty-green and very rough both ways upon the upper surface, scarcely less beneath, rough and hairy along the ribs; sweet-scented when dried; outline oblong, ovate-oblong, or oval, doubly serrate; apex acuminate; base more or less heart-shaped or obtuse, inequilateral; leafstalk short, rough, hairy; stipules small, soon falling.

Inflorescence.—March to April. Preceding the leaves, from the lateral buds of the preceding season, in clusters of nearly sessile, purplish flowers; sterile, fertile, and perfect on the same tree; calyx

5-9-lobed, downy; corolla none; stamens 5-9, anthers dark red; ovary flattened; styles two, purple, downy.

Fruit.—A samara, winged all round, 3/4 inch in diameter, roundish, pubescent over the seed, not fringed, larger than the fruit of *U. Americana*.

Horticultural Value.—Hardy throughout New England; does well in various situations, but prefers a light, sandy or gravelly soil near running water; grows more rapidly than *U. Americana*, and is less liable to the attacks of insects; its large foliage and graceful outline make it worthy of a place in ornamental plantations. Propagated from seed.

Plate XLIX. — Ulmus fulva.

1. Winter buds.
2. Flowering branch,
3. Flower, top view.
4. Flower, side view, part of perianth and stamens removed.
5. Pistil.
6. Fruiting branch.

[Pg 99]

Ulmus racemosa, Thomas.

Cork Elm. Rock Elm.

Habitat and Range. — Dry, gravelly soils, rich soils, river banks.

Quebec through Ontario.

Maine, — not reported; New Hampshire, — rare and extremely local; Meriden and one or two other places (Jessup); Vermont, — rare, Bennington, Pownal (Robbins), Knowlton (Brainerd), Highgate (Eggleston); comparatively abundant in Champlain valley and westward (T. H. Haskins, *Garden and Forest,* V, 86); Massachusetts, — rare; Rhode Island and Connecticut, — not reported native.

South to Tennessee; west to Minnesota, Iowa, Nebraska and Missouri.

Habit. — A large tree, scarcely inferior at its best to *U. Americana,* 50-75 feet high, with a trunk diameter of 2-3 feet; reaching in southern Michigan a height of 100 feet and a diameter of 5 feet; trunk rather slender; branches short and stout, often twiggy in the interior of the tree; branchlets slender, spreading, sometimes with a drooping tendency; head rather narrow, round-topped.

Bark. — Bark of trunk brownish-gray, in old trees irregularly separated into deep, wide, flat-topped ridges; branches grayish-brown; leaf-scars conspicuous; season's shoots light brown, more or less pubescent or glabrous, oblong-dotted; branches and branchlets often marked lengthwise with corky, wing-like ridges.

Winter Buds and Leaves. — Buds ovate to oblong, pointed, scales downy-ciliate, pubescent. Leaves simple, alternate, 3-4 inches long, half as wide, glabrous above, minutely pubescent beneath; outline ovate, doubly serrate (less sharp than the serratures in *U. Americana*); apex acuminate; base inequilateral, produced and rounded on one side, acute or slightly rounded on the other; veins straight; leafstalk short, stout; stipules soon falling. [Pg 100]

Inflorescence. — April to May. Appearing before the leaves from lateral buds of the preceding season, in drooping racemes; calyx lobes 7-8, broad-triangular, with rounded edges and a mostly obtuse apex: pedicels thread-like, jointed; stamens 5-10, exserted, anthers purple, ovary 2-styled: stigmas recurved or spreading.

Fruit. — Samara ovate, broadly oval, or obovate, pubescent, margin densely fringed, resembling fruit of *U. Americana* but somewhat larger.

Horticultural Value. — Hardy throughout New England; prefers a moist, rich soil, in open situations; less variable in habit than the American elm and a smaller tree with smaller foliage, scarcely varying enough to justify its extensive use as a substitute. Not often obtainable in nurseries, but readily transplanted, and easily propagated from the seed.

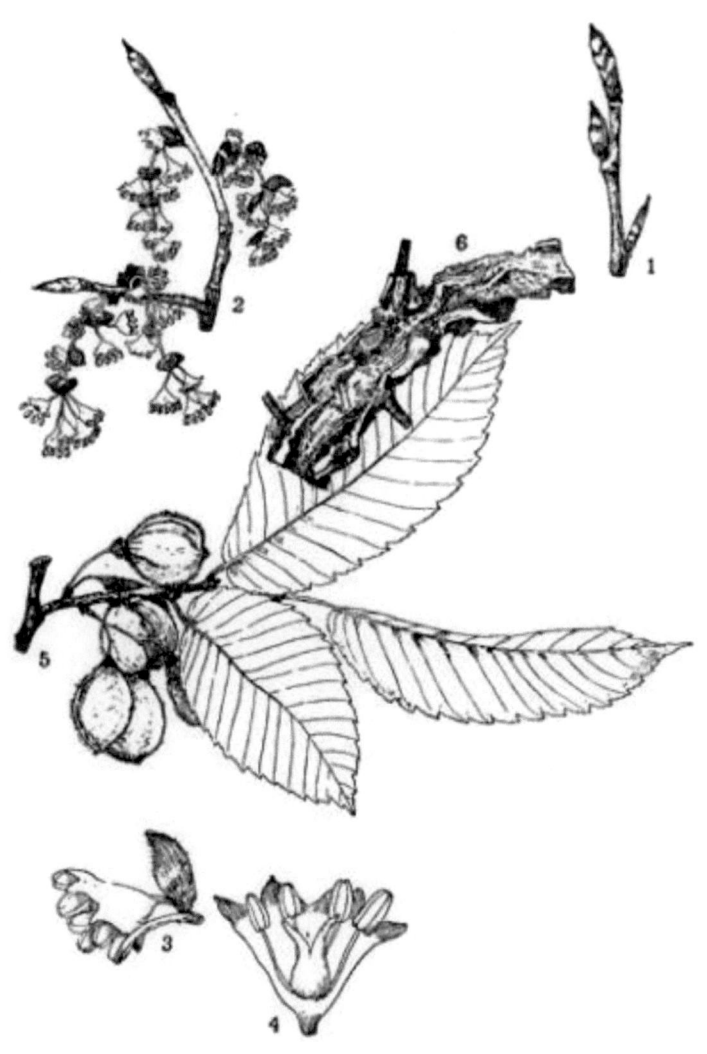

Plate L. — Ulmus racemosa.

1. Winter buds, at the time the flowers open.
2. Flowering branch.
3. Flower, side view.
4. Flower, side view, perianth and stamens partly removed.
5. Fruiting branch.

Celtis occidentalis, L.

Hackberry. Nettle Tree. Hoop Ash. Sugar Berry.

Habitat and Range. — In divers situations and soils; woods, river banks, near salt marshes.

Province of Quebec to Lake of the Woods, occasional.

Maine, — not reported; New Hampshire, — sparingly along the Connecticut valley, as far as Wells river; Vermont, — along Lake Champlain, not common; Norwich and Windsor on the Connecticut (Eggleston); Massachusetts, — occasional throughout the state; Rhode Island, — common (Bailey); Connecticut, — common (J. N. Bishop).

South to the Gulf states; west to Minnesota and Missouri.

[Pg 101]

Habit. — A small or medium-sized tree, 20-45 feet high, with a trunk diameter of 8 inches to 2 feet; attaining farther south a maximum of 100 feet in height, with a trunk diameter of 4-6 feet; variable; most commonly the rough, straight trunk, sometimes buttressed at the base, branches a few feet from the ground, sending out a few large limbs and numerous slender, horizontal or slightly drooping and more or less tortuous branches; head wide-spreading, flattish or often rounded, with deep green foliage which lasts into late autumn with little change in color, and with cherry-like fruit which holds on till the next spring.

Bark. — Bark of trunk in young trees grayish, rough, unbroken, in old trees with deep, short ridges; main branches corrugated; secondary branches close and even; branchlets pubescent; season's shoots reddish-brown, often downy, more or less shining.

Winter Buds and Leaves. — Buds small, ovate, acute, scales chestnut brown. Leaves simple, alternate, extremely variable in size, outline, and texture, usually 2-4 inches long, two-thirds as wide, thin, deep green, and scarcely rough above, more or less pubescent beneath, with numerous and prominent veins, outline ovate to ovate-lanceolate, sharply serrate above the lower third; apex usually narrowly and sharply acuminate; base acutish, inequilateral, 3-nerved, entire; leafstalk slender; stipules lanceolate, soon falling.

Inflorescence. — May. Appearing with the leaves from the axils of the season's shoots, sterile and fertile flowers usually separate on the same tree; flowers slender-stemmed, the sterile in clusters at the base of the shoot, the fertile in the axils above, usually solitary; calyx greenish, segments oblong; stamens 4-6, in the fertile flowers about the length of the 4 lobes, in the sterile exserted; ovary with two long, recurved stigmas.

Fruit. — Drupes, on long slender stems, globular, about the size of the fruit of the wild red cherry, purplish-red when ripe, thin-meated, edible, lasting through the winter.

Horticultural Value. — Hardy throughout New England; grows in all well-drained soils, but prefers a deep, rich, [Pg 102] moist loam. Young trees grow rather slowly and are more or less distorted, and trees of the same age often vary considerably in size and habit; hence it is not a desirable street tree, but it appears well in ornamental grounds. A disease which seriously disfigures the tree is extending to New England, and the leaves are sometimes attacked by insects. Occasionally offered by nurserymen and easily transplanted.

279

Plate LI. — Celtis occidentalis.

1. Winter buds.
2. Flowering branch.
3. Sterile flower.
4. Fertile flower.
5. Fruiting branch.

MORACEÆ. MULBERRY FAMILY.

Morus rubra, L.

Mulberry.

Habitat and Range. — Banks of rivers, rich woods.

Canadian shore of Lake Erie.

A rare tree in New England. Maine, — doubtfully reported; New Hampshire, — Pemigewasset valley, White mountains (Matthews); Vermont, — northern extremity of Lake Champlain, banks of the Connecticut (Flagg), Pownal (Oakes), North Pownal (Eggleston); Massachusetts, — rare; Rhode Island, — no station reported; Connecticut, — rare; Bristol, Plainville, North Guilford, East Rock and Norwich (J. N. Bishop).

South to Florida; west to Michigan, South Dakota, and Texas.

Habit. — A small tree, 15-25 feet in height, with a trunk diameter of 8-15 inches; attaining much greater dimensions in the Ohio and Mississippi basins; a wide-branching, rounded tree, characterized by a milky sap, rather dense foliage, and fruit closely resembling in shape that of the high blackberry.

Bark. — Trunk light brown, rough, and more or less furrowed according to age; larger branches light greenish-brown; season's shoots gray and somewhat downy. [Pg 103]

Winter Buds and Leaves. — Buds ovate, obtuse. Leaves simple, alternate, 4-8 inches long, two-thirds as wide, rough above, yellowish-green and densely pubescent when young; at maturity dark green and downy beneath, turning yellow in autumn; conspicuously reticulated; outline variable, ovate, obovate, oblong or broadly oval, serrate-dentate with equal teeth, or irregularly 3-7-lobed; apex acuminate; base heart-shaped to truncate; stalk 1-2 inches long; stipules linear, serrate, soon falling.

Inflorescence. — May. Appearing with the leaves from the season's shoots, in axillary spikes, sterile and fertile flowers sometimes on the same tree, sometimes on different trees, — sterile flowers in spreading or pendulous spikes, about 1 inch long; calyx 4-parted; petals none; stamens 4, the inflexed filaments of which suddenly straighten themselves as the flower expands: fertile spikes spreading or pendent; calyx 4-parted, becoming fleshy in fruit; ovary sessile; stigmas 2, spreading.

Fruit. — July to August. In drooping spikes about 1 inch long and ½ inch in diameter; dark purplish-red, oblong, sweet and edible; apparently a simple fruit but really made up of the thickened calyx lobes of the spike.

Horticultural Value. — Hardy in southern New England; grows rapidly in a good, moist soil in sun or shade; the large leaves start late and drop early; useful where it is hardy, in low tree plantations or as an undergrowth in woods; readily transplanted, but seldom offered for sale by nurserymen or collectors; propagated from seed.

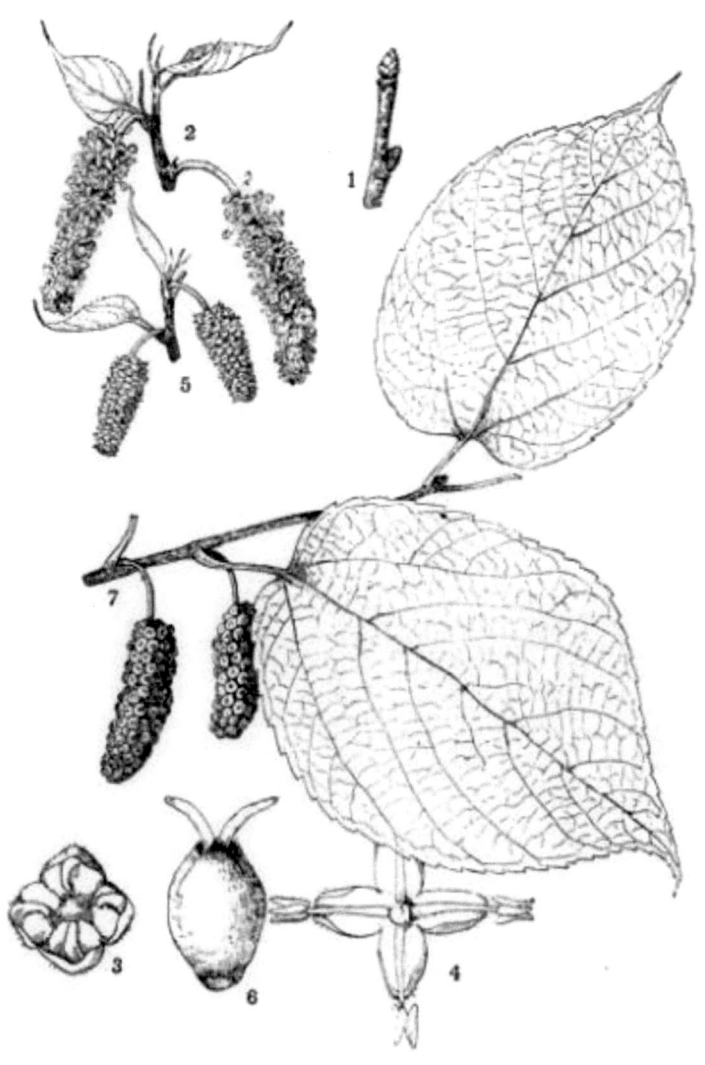

Plate LII. — Morus rubra.

1. Winter buds.
2. Branch with sterile flowers.
3. Sterile flower with stamens incurved.
4. Sterile flower expanded.
5. Branch with fertile flowers.
6. Fertile flower, side view.
7. Fruiting branch.

[Pg 104]

Morus alba, L.

Probably a native of China, where its leaves have from time immemorial furnished food for silkworms; extensively introduced and naturalized in India and central and southern Europe; introduced likewise into the United States and Canada from Ontario to Florida; occasionally spontaneous near dwellings, old trees sometimes marking the sites of houses that have long since disappeared.

It may be distinguished from *M. rubra* by its smooth, shining leaves, its whitish or pinkish fruit, and its greater susceptibility to frost.

MAGNOLIACEÆ. MAGNOLIA FAMILY.

Liriodendron Tulipifera, L.

Tulip Tree. Whitewood. Poplar.

Habitat and Range. — Prefers a rich, loamy, moist soil.

Vermont, — valley of the Hoosac river in the southwestern corner of the state; Massachusetts, — frequent in the Connecticut river valley and westward; reported as far east as Douglas, southeastern corner of Worcester county (R. M. Harper, *Rhodora*, II, 122); Rhode

Island and Connecticut,—frequent, especially in the central and southern portions of the latter state.

South to the Gulf states; west to Wisconsin; occasional in the eastern sections of Missouri and Arkansas; attains great size in the basins of the Ohio and its tributaries, and southward along the Mississippi river bottoms.

Habit.—A medium-sized tree, 50-70 feet high; trunk 2-3 feet in diameter, straight, cylindrical; head rather open, more or less cone-shaped, in the dense forest lifted high and spreading; branches small for the size of the tree, set at varying angles, often decurrent, becoming scraggly with age. The shapely trunk, erect, showy blossoms, green, cone-like fruit, and conspicuous bright green truncate leaves give the tulip tree an air of peculiar distinction. [Pg 105]

Bark.—Bark of trunk ashen-gray and smoothish in young trees, becoming at length dark, seamed, and furrowed; the older branches gray; the season's shoots of a shining chestnut, with minute dots and conspicuous leaf-scars; glabrous or dusty-pubescent; bark of roots pale brown, fleshy, with an agreeable aromatic smell and pungent taste.

Winter Buds and Leaves.—Terminal buds ½-1 inch long; narrow-oblong; flattish; covered by two chestnut-brown dotted scales, which persist as appendages at the base of the leafstalk, often enclosing several leaves which develop one after the other. Leaves simple, alternate, lobed; 3-5 inches long and nearly as broad, dark green and smooth on the upper surface, lighter, with minute dusty pubescence beneath, becoming yellow and russet brown in autumn; usually with four rounded or pointed lobes, the two upper abruptly cut off at the apex, and separated by a slight indentation or notch more or less broad and shallow at the top; all the lobes entire, or 2-3 sublobed, or coarsely toothed; base truncate, acute or heart-shaped; leafstalks as long or longer than the blade, slender, enlarged at the base; stipules 1-2 inches long, pale yellow, oblong, often persisting till the leaf is fully developed.

Inflorescence.—Late May or early June. Flowers conspicuous, solitary, terminal, held erect by a stout stem, tulip-shaped, 1½-2 inches long, opening at the top about 2 inches. There are two triangular bracts which fall as the flower opens; three greenish, concave sepals, at length reflexed; six greenish-yellow petals with an orange spot near the base of each; numerous stamens somewhat shorter than the petals; and pistils clinging together about a central axis.

Fruit. — Cone-like, formed of numerous carpels, often abortive, which fall away from the axis at maturity; each long, flat carpel encloses in the cavity at its base one or two orange seeds which hang out for a time on flexible, silk-like threads.

Horticultural Value. — An ornamental tree of great merit; hardy except in the coldest parts of New England; difficult to transplant, but growing rapidly when established; comes into leaf rather early and holds its foliage till mid-fall, shedding it in a short time when mature; adapts itself readily [Pg 106] to good, light soils, but grows best in moist loam. It has few disfiguring insect enemies. Mostly propagated by seed, but sometimes successfully collected; for sale in the leading nurseries and usually obtainable in large quantities. Of abnormal forms offered by nurserymen, one has an upright habit approaching that of the Lombardy poplar; another has variegated leaves, and another leaves without lobes.

Plate LIII.—Liriodendron Tulipifera.

1. Winter bud, terminal.
2. Opening leaf-bud with stipules.
3. Flowering branch.
4. Fruit.
5. Fruit with many carpels removed.
6. Carpel with seeds.

LAURACEÆ. LAUREL FAMILY.

Sassafras officinale, Nees.

Sassafras Sassafras, Karst.

Sassafras.

Habitat and Range.—In various soils and situations; sandy or rich woods, along the borders of peaty swamps.

Provinces of Quebec and Ontario.

Maine,—this tree grows not beyond Black Point (Scarboro, Cumberland county) eastward (Josselyn's *New England Rarities*, 1672); not reported again by botanists for more than two hundred years; rediscovered at Wells in 1895 (Walter Deane) and North Berwick in 1896 (J. C. Parlin); New Hampshire,—lower Merrimac valley, eastward to the coast and along the Connecticut valley to Bellows Falls; Vermont,—occasional south of the center; Pownal (Robbins, Eggleston); Hartland and Brattleboro (Bates), Vernon (Grant); Massachusetts,—common especially in the eastern sections; Rhode Island and Connecticut,—common.

South to Florida; west to Michigan, Iowa, Kansas, and Texas.

Habit.—Generally a shrub or small tree but sometimes reaching a height of 40-50 feet and a trunk diameter of 2-4 [Pg 107] feet; attaining a maximum in the southern and southwestern states of 80-100 feet in height and a trunk diameter of 6-7 feet; head open, flattish or rounded; branches at varying angles, stout, crooked, and irregular; spray bushy; marked in winter by the contrasting reddish-brown of the trunk, the bright yellowish-green of the shoots and the prominent flower-buds, in early spring by the drooping racemes of yellow flowers, in autumn by the rich yellow or red-tinted foliage and handsome fruit, at all seasons by the aromatic odor and spicy flavor of all parts of the tree, especially the bark of the root.

Bark.—Bark of trunk deep reddish-brown, deeply and firmly ridged in old trees, in young trees greenish-gray, finely and irregularly striate, the outer layer often curiously splitting, resembling a sort of filagree work; branchlets reddish-brown, marked with warts of russet brown; season's shoots at first minutely pubescent, in the fall more or less mottled, bright yellowish-green.

Winter Buds and Leaves.—Flower-buds conspicuous, terminal, ovate to elliptical, the outer scales rather loose, more or less pubescent, the inner glossy, pubescent; lateral buds much smaller. Leaves simple, alternate, often opposite, 3-5 inches long, two-thirds as wide, downy-tomentose when young, at maturity smooth, yellowish-green above, lighter beneath, with midrib conspicuous and minutely hairy; outline of two forms, one oval to oblong, entire, usually rounded at the apex, wedge-shaped at base; the other oval to obovate, mitten-shaped or 3-lobed to about the center, with rounded sinuses; apex obtuse or rounded; base wedge-shaped; leafstalk about 1 inch long; stipules none.

Inflorescence. — April or early May. Appearing with the leaves in slender, bracted, greenish-yellow, corymbous racemes, from terminal buds of the preceding season, sterile and fertile flowers on separate trees, — sterile flowers with 9 stamens, each of the three inner with two stalked orange-colored glands, anthers 4-celled, ovary abortive or wanting: fertile flowers with 6 rudimentary stamens in one row; ovary ovoid; style short.

Fruit. — Generally scanty, drupes, ovoid, deep blue, with club-shaped, bright red stalk. [Pg 108]

Horticultural Value. — Hardy throughout New England; adapted to a great variety of soils, but prefers a stony, well-drained loam or gravel. Its irregular masses of foliage, which color so brilliantly in the fall, make it an extremely interesting tree in plantations, but it has always been rare in nurseries and difficult to transplant; suckers, however, can be moved readily. Propagated easily from seed.

Plate LIV. — Sassafras officinale.

1. Winter buds.
2. Branch with sterile flowers.
3. Sterile flower.
4. Branch with fertile flowers.
5. Fertile flower.
6. Fruiting branch.

HAMAMELIDACEÆ. WITCH HAZEL FAMILY.

Liquidambar Styraciflua, L.

Sweet Gum.

Habitat and Range. — Low, wet soil, swamps, moist woods.

Connecticut, — restricted to the southwest corner of the state, not far from the seacoast; Darien to Five Mile river, probably the northeastern limit of its natural growth.

South to Florida; west to Missouri and Texas.

Habit. — Tree 40-60 feet high, with a trunk diameter of 10 inches to 2 feet, attaining a height of 150 feet and a diameter of 3-5 feet in the Ohio and Mississippi valleys; trunk tall and straight; branches rather small for the diameter and height of the tree, the lower mostly horizontal or declining; branchlets beset with numerous short, rather stout, curved twigs; head wide-spreading, ovoid or narrow-pyramidal, symmetrical; conspicuous in summer by its deep green, shining foliage, in autumn by the splendor of its coloring, and in winter by the long-stemmed, globular fruit, which does not fall till spring.

Bark. — Trunk gray or grayish-brown, in old trees deeply furrowed and broken up into rather small, thickish, loose scales; branches brown-gray; branchlets with or without [Pg 109] prominent corky ridges on the upper side; young twigs yellowish.

Winter Buds and Leaves.—Buds ovate, reddish-brown, glossy, acute. Leaves simple, alternate, regular, 3-4 inches in diameter, dark green turning to reds, purples, and yellows in autumn, cut into the figure of a star by 5-7 equal, pointed lobes, glandular-serrate, smooth, shining on the upper surface, fragrant when bruised; base more or less heart-shaped; stalk slender.

Inflorescence.—May. Developing from a bud of the season; sterile flowers in an erect or spreading, cylindrical catkin; calyx none; petals none, stamens many, intermixed with minute scales: fertile flowers numerous, gathered in a long peduncled head; calyx consisting of fine scales; corolla none; pistil with 2-celled ovary and 2 long styles.

Fruit.—In spherical, woody heads, about 1 inch in diameter, suspended by a slender thread: a sort of aggregate fruit made up of the hardened, coherent ovaries, holding on till spring, each containing one or two perfect seeds.

Horticultural Value.—Hardy along the southern shores of New England; grows in good wet or dry soils, preferring clays. Young plants are tender in Massachusetts, but if protected a few seasons until well established make hardy trees of medium size. It is offered by nurserymen, but must be frequently transplanted to be moved with safety; rate of growth rather slow and nearly uniform to maturity. Propagated from seed.

Plate LV.—Liquidambar styraciflua.

1. Winter buds.
2. Flowering branch.
3. Sterile flower.
4. Fertile flower.
5. Fruiting branch.

[Pg 110]

PLATANACEÆ. PLANE-TREE FAMILY.

Platanus occidentalis, L.

Buttonwood. Sycamore. Buttonball. Plane Tree.

Habitat and Range.—Near streams, river bottoms, and low, damp woods.

Ontario.

Maine,—apparently restricted to York county; New Hampshire,—Merrimac valley towards the coast; along the Connecticut as far as Walpole; Vermont,—scattering along the river shores, quite abundant along the Hoosac in Pownal (Eggleston); Massachusetts,—occasional; Rhode Island and Connecticut,—rather common.

South to Florida; west to Minnesota, Nebraska, Kansas, and Texas.

Habit.—A tree of the first magnitude, 50-100 feet and upwards in height, with a diameter of 3-8 feet; reaching in the rich alluvium of the Ohio and Mississippi valleys a maximum of 125 feet in height and a diameter of 20 feet; the largest tree of the New England forest, conspicuous by its great height, massive trunk and branches, and by its magnificent, wide-spreading, dome-shaped or pyramidal, open head. The sunlight, streaming through the large-leafed, rusty foliage, reveals the curiously mottled patchwork bark; and the long-stemmed, globular fruit swings to every breeze till spring comes again.

The lower branches are often very long and almost horizontal, and the branchlets frequently have a tufted, broom-like appearance, due probably to the action of a fungous disease on the young growth.

Bark.—Bark of trunk and large branches dark greenish-gray, sometimes rough and closely adherent, but usually flaking off in broad, thin, brittle scales, exposing the green or buff inner bark, which becomes nearly white on exposure; branchlets light brown, sometimes ridgy towards the ends, marked with numerous inconspicuous dots.

Winter Buds and Leaves.—Buds short, ovate, obtuse, enclosed in the swollen base of a petiole, and, after the fall of the leaf, [Pg 111] encircled by the leaf-scar. Leaves simple, alternate, 5-6 inches long, 7-10 wide, pubescent on both sides when young, at maturity light rusty-green above, light green beneath, finally smooth, turning yellow in autumn, coriaceous; outline reniform; margin coarse-toothed or sinuate-lobed, the short lobes ending in a sharp point; base heart-shaped to nearly truncate; leafstalk 1-2 inches long, swollen at the base; stipules sheathing, often united, forming a sort of ruffle.

Inflorescence.—May. In crowded spherical heads; flowers of both kinds with insignificant calyx and corolla,—sterile heads from terminal or lateral buds of the preceding season, on short and pendulous stems; stamens few, usually 4, anthers 2-celled: fertile heads from shoots of the season, on long, slender stems, made up of closely compacted ovate ovaries with intermingled scales, ovaries surmounted by hairy one-sided recurved styles, with bright red stigmas.

Fruit.—In heads, mostly solitary, about 1 inch in diameter, persistent till spring: nutlets small, hairy, 1-seeded.

Horticultural Value.—Hardy throughout New England; prefers a deep, rich, loamy soil near water, but grows in almost any situation; of more rapid growth than almost any other native tree, and formerly planted freely in ornamental grounds and on streets, but fungous diseases disfigure it so seriously, and the late frosts so often kill the young leaves that it is now seldom obtainable in nurseries; usually propagated from seed. The European plane, now largely grown in some nurseries, is a suitable substitute.

Plate LVI.—Platanus occidentalis.

1. Winter buds.
2. Flowering branch with sterile and fertile heads.
3. Stamen.
4. Pistil.
5. Fruiting branch.
6. Stipule.
7. Bud with enclosing base of leafstalk.

[Pg 112]

POMACEÆ. APPLE FAMILY.

Trees or shrubs; leaves simple or pinnate, mostly alternate, with stipules free from the leafstalk and usually soon falling; flowers regular, perfect; calyx 5-lobed; calyx-tube adnate to ovary; petals 5, inserted on the disk which lines the calyx-tube; stamens usually many, distinct, inserted with the petals; carpels of the ovary 1-5, partially or entirely united with each other; ovules 1-2 in each carpel; styles 1-5; fruit a fleshy pome, often berry-like or drupe-like, formed by consolidation of the carpels with the calyx-tube.

Pyrus. Malus. Amelanchier. Cratægus.

Pyrus Americana, DC.

Sorbus Americana, Marsh.

Mountain Ash.

Habitat and Range.—River banks, cool woods, swamps, and mountains.

Newfoundland to Manitoba.

Maine,—common; New Hampshire,—common along the watersheds of the Connecticut and Merrimac rivers and on the slopes of the White mountains; Vermont,—abundant far up the slopes of the Green mountains; Massachusetts,—Graylock, Wachusett, Watatic, and other mountainous regions; rare eastward; Rhode Island and Connecticut,—occasional in the northern sections.

South, in cold swamps and along the mountains to North Carolina; west to Michigan and Minnesota.

Habit.—A small tree, 15-20 feet high, often attaining in the woods of northern Maine and on the slopes of the White mountains a height of 25-30 feet, with a trunk diameter of 12-15 inches; reduced at its extreme altitudes to a low shrub; head, in open ground, pyramidal or roundish; branches spreading and slender. [Pg 113]

Bark.—Closely resembling bark of *P. sambucifolia*.

Winter Buds and Leaves.,—Buds more or less scythe-shaped, acute, smooth, glutinous. Leaves pinnately compound, alternate; stem grooved, enlarged at base, reddish-brown above; stipules deciduous; leaflets 11-19, 2-4 inches long, bright green above, paler beneath, smooth, narrow-oblong or lanceolate, the terminal often elliptical, finely and sharply serrate above the base; apex acuminate; base roundish to acute and unequally sided; sessile or nearly so, except in the odd leaflet.

Inflorescence.—In terminal, densely compound, large and flattish cymes; calyx 5-lobed; petals 5, white, roundish, short-clawed; stamens numerous; ovary inferior; styles 3.

Fruit.—Round, bright red, about the size of a pea, lasting into winter.

Horticultural Value.—Hardy throughout New England; prefers a good, well-drained soil; rate of growth slow and nearly uniform. It is readily transplanted and would be useful on the borders of woods, in plantations of low trees, and in seaside exposures. Rare in nurseries and seldom for sale by collectors. The readily obtainable and more showy European *P. aucuparia* is to be preferred for ornamental purposes.

Plate LVII. — Pyrus Americana.

1. Winter buds.
2. Flowering branch.
3. Flower with part of perianth and stamens removed.
4. Petal.
5. Fruiting branch.

Pyrus sambucifolia, Cham. & Schlecht.

Sorbus sambucifolia, Rœm.

Mountain Ash.

Habitat and Range.—Mountain slopes, cool woods, along the shores of rivers and ponds, often associated with *P. Americana*, but climbing higher up the mountains.

From Labrador and Nova Scotia west to the Rocky mountains, then northward along the mountain ranges to Alaska. [Pg 114]

Maine,—abundant in Aroostook county, Piscataquis county, Somerset county at least north to the Moose river, along the boundary mountains, about the Rangeley lakes and locally on Mount Desert Island; New Hampshire,—in the White mountain region; Vermont,—Mt. Mansfield, Willoughby mountain (Pringle); undoubtedly in other sections of these states; to be looked for along the edges of deep, cool swamps and at considerable elevations.

South of New England, probably only as an escape from cultivation; west through the northern tier of states to the Rocky mountains, thence northward along the mountain ranges to Alaska and south to New Mexico and California.

Habit.—A shrub 3-10 feet high, or small tree rising to a height of 15-25 feet, reaching its maximum in northern New England, where it occasionally attains a height of 30-35 feet, with a trunk diameter of 15 inches. It forms an open, wide-spreading, pyramidal or roundish head, resembling the preceding species in the color of bark, in foliage and fruit. Whether these are two distinct species is at the present problematical, as there are many intermediate forms, and the same tree sometimes furnishes specimens that would indubitably be referred to different species.

Bark.—On old trees light brown and roughish on the trunk, separating into small scales curling up on one side; large limbs light-colored, smoothish, often conspicuously marked with coarse horizontal blotches and leaf-scars; season's shoots light brown, smooth, silvery dotted.

Winter Buds and Leaves.—Terminal bud 1 inch long, lateral ½ inch, appressed, brownish, scythe-shaped, acute, more or less glutinous. Leaves pinnately compound, alternate, stems grooved and reddish above, enlarged at base; stipules deciduous; leaflets 7-15, the odd one stalked, 1-3 inches long, ½-1 inch wide, bright green above, paler beneath, smooth, mostly ovate-oblong, serrate above the base; apex rounded or more usually tapering suddenly to a short point, or rarely acuminate; base inequilateral. [Pg 115]

Inflorescence. — In broad, compound cymes at the ends of the branches; flowers white and rather larger than those of *P. Americanus*; calyx 5-lobed; petals 5, ovate, short-clawed; stamens numerous; pistil 3-styled.

Fruit. — In broad cymes; berries bright red, roundish, rather larger than those of *P. Americana*, holding on till winter.

Horticultural Value. — Hardy in New England, though of shrublike proportions in the southern sections; grows in exposed situations inland, and along the seashore. The dwarf habit, graceful foliage, and showy fruit give it an especial value in artificial plantations; but it is seldom for sale in nurseries and only occasionally by collectors. It is readily transplanted and is propagated by seed.

Note. — In the European mountain ash, *P. aucuparia*, the leaves have a blunter apex than is usually found in either of the American species, and have a more decided tendency to double serration.

Plate LVIII. — Pyrus sambucifolia.

1. Winter buds.
2. Flowering branch.
3. Flower with part of perianth and stamens removed.
4. Fruiting branch.

Pyrus communis, L.

Pear Tree.

The common pear, introduced from Europe; a frequent escape from cultivation throughout New England and elsewhere; becomes scraggly and shrubby in a wild state.

Pyrus Malus, L.

Malus Malus, Britton.

Apple Tree.

The common apple; introduced from Europe; a more or less frequent escape wherever extensively cultivated, like the pear showing a tendency in a wild state to reversion. [Pg 116]

Amelanchier Canadensis, Medic.

Shadbush. June-berry.

Habitat and Range.—Dry, open woods, hillsides.

Newfoundland and Nova Scotia to Lake Superior.

New England,—throughout.

South to the Gulf of Mexico; west to Minnesota, Kansas, and Louisiana.

Habit.—Shrub or small tree, 10-25 feet high, with a trunk diameter of 6-10 inches, reaching sometimes a height of 40 feet and trunk diameter of 18 inches; head rather wide-spreading, slender-branched, open; conspicuous in early spring, while other trees are yet naked, by its profuse display of loose spreading clusters of white flowers, and the delicate tints of the silky opening foliage.

Bark.—Trunk and large branches greenish-gray, smooth; branchlets purplish-brown, smooth.

Winter Buds and Leaves.—Buds small, oblong-conical, pointed. Leaves 2-3-½ inches long, about half as wide, slightly pubescent when young, dark bluish-green above at maturity, lighter beneath; outline varying from ovate to obovate, finely and sharply serrate; apex pointed or mucronate, often abruptly so; base somewhat heart-shaped or rounded; leafstalk about 1 inch long; stipules slender, silky, ciliate, soon falling.

Inflorescence.—April to May. Appearing with the leaves at the end of the branchlets in long, loose, spreading or drooping, nearly glabrous racemes; flowers large; calyx 5-cleft, campanulate, pubescent to nearly glabrous; segments lanceolate, acute, reflexed; petals

5, whole, narrow-oblong or oblong-spatulate, about 1 inch long, two to three times the length of the calyx; stamens numerous: ovary with style deeply 5-parted.

Fruit.—June to July. In drooping racemes, globose, passing through various colors to reddish, purplish, or black purple, long-stemmed, sweet and edible without decided flavor. [Pg 117]

Horticultural Value.—Hardy throughout New England; grows in all soils and situations except in wet lands, but prefers deep, rich, moist loam; very irregular in its habit of growth, sometimes forming a shrub, at other times a slender, unsymmetrical tree, and again a symmetrical tree with well-defined trunk. Its beautiful flowers, clean growth, attractive fruit and autumn foliage make it a desirable plant in landscape plantations where it can be grouped with other trees. Occasionally in nurseries; procurable from collectors.

Plate LIX. — Amelanchier Canadensis.

1. Winter buds.
2. Flowering branch.
3. Flower with part of perianth and stamens removed.
4. Fruiting branch.

Cratægus.

A revision of genus *Cratægus* has long been a desideratum with botanists. The present year has added numerous new species, most of which must be regarded as provisional until sufficient time has elapsed to note more carefully the limits of variation in previously existing species and to eliminate possible hybrids. During the present period of uncertainty it seems best to exclude most of the new species from the manuals until their status has been satisfactorily established by raising plants from the seed, or by prolonged observation over wide areas.

Cratægus Crus-Galli, L.

Cockspur Thorn.

Rich soils, edge of swamps.

Quebec to Manitoba.

Found sparingly in western Vermont (*Flora of Vermont*, 1900); southern Connecticut (C. H. Bissell).

South to Georgia; west to Iowa.

A small tree, 10-25 feet in height and 6-12 inches in trunk diameter; best distinguished by its thorns and leaves. [Pg 118]

Thorns numerous, straight, long (2-4 inches), slender; leaves thick, smooth, dark green, shining on the upper surface, pale beneath, turning dark orange red in autumn; outline obovate-oblanceolate, serrate above, entire or nearly so near base; apex acute or rounded; base decidedly wedge-shaped shaped; leafstalks short.

Fruit globose or very slightly pear-shaped, remaining on the tree throughout the winter.

Hardy throughout southern New England; used frequently for a hedge plant.

Cratægus punctata, Jacq.

Thickets, hillsides, borders of forests.

Quebec and Ontario.

Small tree, common in Vermont (Brainerd) and occasional in the other New England states.

South to Georgia.

Thorns 1-2 inches long, sometimes branched; leaves 1-2½ inches long, smooth on the upper surface, finally smooth and dull beneath; outline obovate, toothed or slightly lobed above, entire or nearly so beneath, short-pointed or somewhat obtuse at the apex, wedge-shaped at base; leafstalk slender, 1-2 inches long; calyx lobes linear, entire; fruit large, red or yellow.

Cratægus coccinea, L.

In view of the fact of great variation in the bark, leaves, inflorescence, and fruit of plants that have all passed in this country as *C. coccinea*, and in view of the further uncertainty as to the plant on which the species was originally founded, it seems "best to consider the specimen in the Linnæan herbarium as the type of *C. coccinea* which can be described as follows:

"Leaves elliptical or on vigorous shoots mostly semiorbicular, acute or acuminate, divided above the middle into numerous acute coarsely glandular-serrate lobes, cuneate and finely glandular-serrate below the middle and often quite entire toward the [Pg 119] base, with slender midribs and remote primary veins arcuate and running to the points of the lobes, at the flowering time membranaceous, coated on the upper surface and along the upper surface of the midribs and veins with short soft white hairs, at maturity thick, coriaceous, dark green and lustrous on the upper surface, paler on the lower surface, glabrous or nearly so, 1½-2 inches long and 1-1½ inches wide, with slender glandular petioles 3/4-1 inch long, slightly grooved on the upper surface, often dark red toward the base, and like the young branchlets villous with pale soft hairs; stipules lanceolate to oblanceolate, conspicuously glandular-serrate with dark red glands, ½-¾4 inch long. Flowers ½-¾ inch in diameter when fully expanded, in broad, many-flowered, compound tomentose cymes; bracts and bractlets linear-lanceolate, coarsely glandular-serrate, caducous; calyx tomentose, the lobes lanceolate, glandular-serrate, nearly glabrous or tomentose, persistent, wide-spreading or erect on the fruit, dark red above at the base; stamens 10; anthers yellow; styles 3 or 4. Fruit subglobose, occasionally rather longer than broad, dark crimson, marked with scattered dark dots, about ½ inch in diameter, with thin, sweet, dry yellow flesh; nutlets 3 or 4, about ¼ inch long, conspicuously ridged on the back with high grooved ridges.

"A low, bushy tree, occasionally 20 feet in height with a short trunk 8-10 inches in diameter, or more frequently shrubby and forming wide dense thickets, and with stout more or less zigzag branches bright chestnut brown and lustrous during their first year, ashy-gray during their second season and armed with many stout, chestnut-brown, straight or curved spines 1-1½ inches long. Flowers late in May. Fruit ripens and falls toward the end of October, usually after the leaves.

"Slopes of hills and the high banks of salt marshes usually in rich, well-drained soil, Essex county, Massachusetts, John Robinson, 1900; Gerrish island, Maine, J. G. Jack, 1899-1900; Brunswick, Maine, Miss Kate Furbish, May, 1899; Newfoundland, A. C. Waghorne, 1894." [1]

[1] Prof. C. S. Sargent in *Bot. Gaz.*, XXXI, 12. By permission of the publishers. [Pg 120]

Cratægus mollis, Scheele.

Cratægus subvillosa, Schr. Cratægus coccinea, **var.** *mollis, T. & G.*

Thorn.

Habitat and Range. — Bordering on low lands and along streams.

Provinces of Quebec and Ontario.

Maine, — as far north as Mattawamkeag on the middle Penobscot, Dover on the Piscataquis, and Orono on the lower Penobscot; reported also from southern sections; Vermont, — Charlotte (Hosford); Massachusetts, — in the eastern part infrequent; no stations reported in the other New England states.

South to Pennsylvania, Louisiana, and Texas; west to Michigan and Missouri.

Habit. — Shrub or often a small tree, 20-30 feet high, with trunk 6-12 inches in diameter, often with numerous suckers; branches at 4-6 feet from the ground, at an acute angle with the stem, lower often horizontal or declining; head spreading, widest at base, spray short, angular, and bushy; thorns slender, 1-3 inches long, straight or slightly recurved.

Bark. — Bark of the whole tree, except the ultimate shoots, light gray, on the trunk and larger branches separating lengthwise into thin narrow plates, in old trees dark gray and more or less shreddy; season's shoots reddish or yellowish-brown, glossy.

Winter Buds and Leaves. — Buds small, ovate, reddish-brown, shining; scales broad, glandular-edged. Leaves simple, alternate, 3-5 inches long, light green above, lighter beneath, broad-ovate to broad-elliptical; rather regularly and slightly incised with fine, glandular-tipped teeth; apex acute; base wedge-shaped, truncate, or subcordate; roughish above and slightly pubescent beneath, especially along the veins; leaf-stalk pubescent; stipules linear, glandular-edged, deciduous. [Pg 121]

Inflorescence. — May to June. In cymes from the season's growth; flowers white, 3/4 inch broad, ill-smelling; calyx lobes 5, often incised, pubescent; petals roundish; stamens indefinite, styles 3-5; flower stems pubescent; bracts glandular.

Fruit. — A drupe-like pome, ½-1 inch long, bright scarlet, larger than the fruit of the other New England species; ripens and falls in September.

Horticultural Value. — Hardy in New England. An attractive and useful tree in low plantations; rarely for sale by nurserymen or collectors; propagated from the seed.

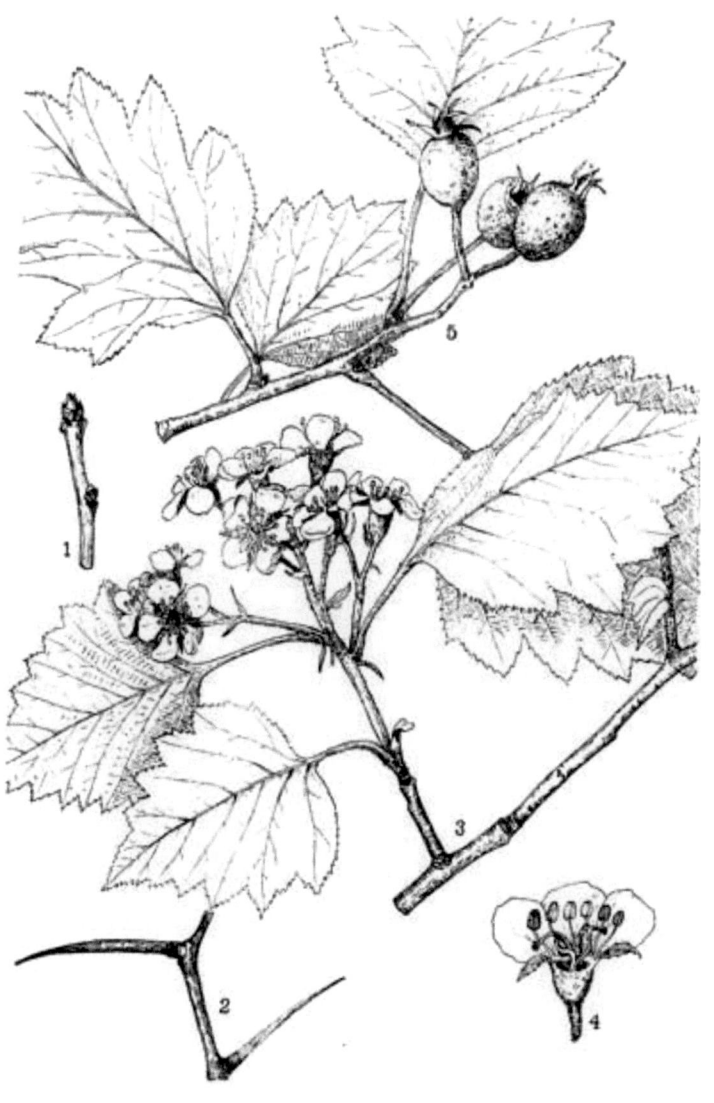

Plate LX.—Cratægus mollis.

1. Winter buds.
2. Branch with thorns.
3. Flowering branch.
4. Flower with part of perianth and stamens removed.
5. Fruiting branch.

Note.—The New England plants here put under the head of *Cratægus mollis* have been referred by Prof. C. S. Sargent to *Cratægus submollis* (*Bot. Gaz.*, XXXI, 7, 1901). The new species differs from the true *Cratægus mollis* in its smaller ovate leaves with cuneate base and more or less winged leafstalk, in the smaller number of its stamens, usually 10, and in its pear-shaped orange-red fruit, which drops in early September.

It is also probable that *C. Arnoldiana*, Sargent, new species, has been collected in Massachusetts as *C. mollis*. It differs from *C. submollis* "in its broader, darker green, more villose leaves which are usually rounded, not cuneate at the base, in its smaller flowers, subglobose, not oblong or pear-shaped, crimson fruit with smaller spreading calyx lobes, borne on shorter peduncles and ripening two or three weeks earlier, and by its much more zigzag and more spiny branches, which make this tree particularly noticeable in winter, when it may readily be recognized from all other thorn trees."—C. S. Sargent in *Bot. Gaz.*, XXXI, 223, 1901.

[Pg 122]

DRUPACEÆ. PLUM FAMILY.

Trees or shrubs; bark exuding gum; bark, leaves, and especially seeds of several species abounding in prussic acid; leaves simple, alternate, mostly serrate; stipules small, soon falling; leafstalk often with one to several glands; flowers in umbels, racemes, or solitary, regular; calyx tube free from the ovary, 5-lobed; petals 5, inserted on the calyx; stamens indefinite, distinct, inserted with the petals; pistil 1, ovary with 1 carpel, 1-seeded; fruit a more or less fleshy drupe.

Prunus nigra, Ait.

Prunus Americana, var. *nigra, Waugh.*

Wild Plum. Red Plum. Horse Plum. Canada Plum.

Habitat and Range.—Native along streams and in thickets, often spontaneous around dwellings and along fences.

From Newfoundland through the valley of the St. Lawrence to Lake Manitoba.

Maine,—abundant in the northern sections and common throughout; New Hampshire and Vermont,—frequent, especially in the northern sections; Massachusetts,—occasional; Rhode Island and Connecticut,—not reported.

Rare south of New England; west to Wisconsin.

Habit.—A shrub or small tree, 20-25 feet high; trunk 5-8 inches in diameter; branches stout, ascending, somewhat angular, with short, rigid branchlets, forming a stiff, narrow head.

Bark.—Bark of trunk grayish-brown, smooth in young trees, in old trees separating into large plates; smaller branches dark brown, season's shoots green.

Winter Buds and Leaves.—Buds small, ovate, acute, dark brown.

Leaves 3-5 inches long, light green on the upper side, paler beneath, pubescent when young; outline ovate-obovate or orbicular, crenulate-serrate; teeth not bristle-tipped; apex [Pg 123] abruptly acuminate; base wedge-shaped, rounded, somewhat heart-shaped, or narrowing to a short petiole more or less red-glandular near the blade; stipules usually linear, ciliate, soon falling.

Inflorescence.—Appearing in May before the leaves, in lateral, 2-3-flowered, slender-stemmed umbels; flowers about an inch broad, white when expanding, turning to pink; calyx 5-lobed, glandular; petals 5, obovate-oblong, contracting to a claw; stamens numerous; style 1, stigma 1.

Fruit.—A drupe, oblong-oval, 1-1½ inches long, orange or orange-red, skin tough, flesh adherent to the flat stone and pleasant to the taste. The fruit toward the southern limit of the species is often abortive, or develops through the growth of a fungus into monstrous forms.

Horticultural Value. — Hardy throughout New England, and will grow, when not shaded, in almost any dry or moist soil. It has a tendency to sucker freely, forming low, broad thickets, especially attractive from their early spring flowers and handsome autumn leaves.

Plate LXI. — Prunus nigra.

1. Winter buds.
2. Flowering branch.
3. Flower with petals removed.
4. Petal.
5. Fruiting branch.
6. Stone.

Prunus Americana, Marsh.

A rare plant in New England, scarcely attaining tree-form. The most northern station yet reported is along the slopes of Graylock, Massachusetts, where a few scattered shrubs were discovered in 1900 (J. R. Churchill). In Connecticut it seems to be native in the vicinity of Southington, shrubs, and small trees 10-15 feet high (C. H. Bissell *in lit.*, 1900); New Milford and Munroe, small trees (C. K. Averill).

Distinguished from *P. nigra* by its sharply toothed leaves, smaller blossoms (the petals of which do not turn pink), and by its globose fruit. [Pg 124]

Plate LXII. — Prunus Americana.

1. Winter buds.
2. Flowering branch.
3. Flower with part of perianth and stamens removed.
4. Petal.
5. Flowering branch.
6. Stone.

Prunus Pennsylvanica, L. f.

Red Cherry. Pin Cherry. Pigeon Cherry. Bird Cherry.

Habitat and Range. — Roadsides, clearings, burnt lands, hill slopes, occasional in rather low grounds.

From Labrador to the Rocky mountains, through British Columbia to the Coast Range.

Throughout New England; very common in the northern portions, as high up as 4500 feet upon Katahdin, less common southward and near the seacoast.

South to North Carolina; west to Minnesota and Missouri.

Habit. — A slender tree, seldom more than 30 feet high; trunk 8-10 inches in diameter, erect; branches at an angle of 45° or less; head rather open, roundish or oblong, characterized in spring by clusters of long-stemmed white flowers, and in autumn by a profusion of small red fruit.

Bark. — Bark of trunk in fully grown trees dark brownish-red, conspicuously marked with coarse horizontal lines; the outer layer peeling off in fine scales, disclosing a brighter red layer beneath; in young trees very smooth and shining throughout; lines very conspicuous in the larger branches; branchlets brownish-red with small horizontal lines; spray and season's shoots polished red, with minute orange dots.

Winter Buds and Leaves. — Buds small, broad-conical, acute. Leaves numerous, 3-4 inches long, 1-2 inches wide, light green and shining on both sides, ovate-lanceolate, oval or [Pg 125] oblong-lanceolate, finely serrate; teeth sharp-pointed, sometimes incurved; apex acuminate; base obtuse or roundish; midrib depressed above; leafstalks short, channeled; stipules falling early.

Inflorescence. — June. Appearing with the leaves, in lateral clusters, the flowers on long, slender, somewhat branching stems; calyx 5-cleft; segments thin, reflexed; petals 5, white, obovate, short-clawed; stamens numerous; pistil 1; style 1.

Fruit. — About the size of a pea, round, light red, thin-meated and sour: stone oval or ovate.

Horticultural Value. — Hardy throughout New England; prefers a light gravelly loam, but grows in poor soils and exposed situations; habit so uncertain and tendency to sprout so decided that it is not wise to use it in ornamental plantations; sometimes very useful in sterile land. A variety with transparent yellowish fruit is occasionally met with, but is not yet in cultivation.

335

Plate LXIII. — Prunus Pennsylvanica.

1. Winter buds.
2. Flowering branch.
3. Flower with part of perianth and stamens removed.
4. Petal.
5. Fruiting branch.

Prunus Virginiana, L.

Chokecherry.

Habitat and Range. — In varying soils; along river banks, on dry plains, in woods, common along walls, often thickets.

From Newfoundland across the continent, as far north on the Mackenzie river as 62°.

Common throughout New England; at an altitude of 4500 feet upon Mt. Katahdin.

South to Georgia; west to Minnesota and Texas.

Habit. — Usually a shrub a few feet high, but occasionally a tree 15-25 feet in height, with a trunk diameter of 5-6 inches; [Pg 126] head, in open places, spreading, somewhat symmetrical, with dull foliage, but very attractive in flower and fruit, the latter variable in color and quantity.

Bark. — Trunk and branches dull gray, darker on older trees, rough with raised buff-orange spots; branchlets dull grayish or reddish brown; season's shoots lighter, minutely dotted. Bitter to the taste.

Winter Buds and Leaves. — Buds 1-1¼ inches long, conical, sharp-pointed, brown, slightly divergent from the stem.

Leaves 2-5 inches long and two-thirds as wide, dull green on the upper side, lighter beneath, obovate or oblong, thin, finely, sharply, and often doubly serrate; apex abruptly pointed; base roundish, obtuse or slightly heart-shaped; leafstalk round, grooved, with two or more glands near base of leaf; stipules long, narrow, ciliate, falling when the leaves expand.

Inflorescence. — Appearing in May, a week earlier than *P. serotina*, terminating lateral, leafy shoots of the season in numerous handsome, erect or spreading racemes, 2-4 inches long; flowers short-stemmed, about ⅓ inch across; petals white, roundish; edge often eroded; calyx 5-cleft with thin reflexed lobes, soon falling; stamens numerous; pistil 1; style 1.

Fruit. — In drooping racemes; varying from yellow to nearly black, commonly bright red, edible, but more or less astringent; stem somewhat persistent after the cherry falls.

Horticultural Value. — Hardy throughout New England; grows in almost any soil, but prefers a deep, rich, moist loam. Vigorous young trees are attractive, but in New England they soon begin to show dead branches, and are so seriously affected by insects and fungous diseases that it is not wise to use them in ornamental plantations, or to permit them to remain on the roadside.

Plate LXIV. — Prunus Virginia.

1. Winter buds.
2. Flowering branch.
3. Flower with part of perianth and stamens removed.
4. A petal.
5. Fruiting branch.

[Pg 127]

Prunus serotina, Ehrh.

Rum Cherry. Black Cherry.

Habitat and Range. — In all sorts of soils and exposures; open places and rich woods.

Nova Scotia to Lake Superior.

Maine, — not reported north of Oldtown (Penobscot county); frequent throughout the other New England states.

South to Florida; west to North Dakota, Kansas, and Texas, extending through Mexico, along the Pacific coast of Central America to Peru.

Habit. — Usually a medium-sized tree, 30-50 feet in height, with a trunk diameter varying from 8 or 10 inches to 2 feet; attaining much greater dimensions in the middle and southern states; branches few, large, often tortuous, subdividing irregularly; head open, widest near the base, rather ungraceful when naked, but very attractive when clothed with bright green, polished foliage, profusely decked with white flowers, or laden with drooping racemes of handsome black fruit.

Bark. — Bark of trunk deep reddish-brown and smooth in young trees, in old trees very rough, separating into close, thick, irregular, blackish scales; branches dark reddish-brown, marked with small oblong, raised dots. Bitter to the taste.

Winter Buds and Leaves. — Buds ovate, ⅛ inch long, covered with imbricated brown scales.

Leaves 2-5 inches long, about half as wide, dark green above and glossy when full grown, paler below, turning in autumn to orange, deep red, or pale yellow, firm, smooth on both sides, elliptical, oblong, or lanceolate-oblong; finely serrate with short, incurved teeth; apex sharp; base acute or roundish; meshes of veins minute; petioles ½ inch long, with usually two or more glands near the base of the leaf; stipules glandular-edged, falling as the leaf expands.

Inflorescence. — May to June. From new leafy shoots, in simple, loose racemes, 4-5 inches long; flowers small; calyx with 5 short teeth separated by shallow sinuses, persistent [Pg 128] after the cherry falls; petals 5, spreading, white, obovate; stamens numerous; pistil one; style single.

Fruit. — September. Somewhat flattened vertically, ¼ inch in diameter; purplish-black, edible, slightly bitter.

Horticultural Value. — Hardy in New England; in rich soil in open situations young trees grow very rapidly, old trees rather slowly. Seldom used for ornamental purposes, but serves well as a nurse tree for forest plantations, or where quick results and a luxurious foliage effect is desired, on inland exposures or near the seacoast. The branches are very liable to disfigurement by the black-knot and the foliage by the tent-caterpillar. Large plants are seldom for sale, but seedlings may be obtained in large quantities and at low prices. A weeping horticultural form is occasionally offered. Propagated from seed.

Plate LXV.—Prunus serotina.

1. Winter buds.
2. Flowering branch.
3. Flower with part of perianth and stamens removed.
4. A petal.
5. Fruiting branch.
6. Mature leaf.

Prunus Avium, L.

Mazard Cherry.

Introduced from England; occasionally spontaneous along fences and the borders of woodlands. As an escape, 25-50 feet high, with a trunk diameter of 1-2 feet; head oblong or ovate; branches mostly ascending. Leaves ovate to obovate, more or less pubescent beneath, serrate, 3-5 inches long; leafstalk about ½ inch long, often glandular near base of leaf; inflorescence in umbels; flowers white, expanding with the leaves; fruit dark red, sweet, mostly inferior or blighted. [Pg 129]

LEGUMINOSÆ. PULSE FAMILY.

Gleditsia triacanthos, L.

Honey Locust. Three-thorned Acacia.

Habitat and Range.—In its native habitat growing in a variety of soils; rich woods, mountain sides, sterile plains.

Southern Ontario.

Maine,—young trees in the southern sections said to have been produced from self-sown seed (M. L. Fernald); New Hampshire and Vermont,—introduced; Massachusetts,—occasional; Rhode Island,—introduced and fully at home (J. F. Collins); Connecticut,—not reported. Probably sparingly naturalized in many other places in New England.

Spreading by seed southward; indigenous along the western slopes of the Alleghanies in Pennsylvania; south to Georgia and Alabama; west from western New York through southern Ontario (Canada) and Michigan to Nebraska, Kansas, Indian territory, and Texas.

Habit.—A medium-sized tree, reaching a height of 40-60 feet and a trunk diameter of 1-3 feet; becoming a tree of the first magnitude in the river bottoms of Ohio, Kentucky, and Tennessee; trunk dark and straight, the upper branches going off at an acute angle, the lower often horizontal, both trunk and larger branches armed above the axils with stout, sharp-pointed, simple, three-pronged or numerously branched thorns, sometimes clustered in forbidding tangles a foot or two in length; head wide-spreading, very open, rounded or flattish, with extremely delicate, fern-like foliage lying in graceful planes or masses; pods flat and pendent, conspicuous in autumn.

Bark.—Trunk and larger branches a sombre iron gray, deepening on old trees almost to black, yellowish-brown in second year's growth; season's shoots green, marked with short buff, longitudinal lines; branchlets rough-dotted.

Winter Buds and Leaves.—Winter buds minute, in clusters of [Pg 130] three or four, the upper the largest. Leaves compound, once to twice pinnate, both forms often in the same leaf, alternate, 6 inches to 1 foot long, rachis abruptly enlarged at base and covering the winter buds: leaflets 18-28, ¾-¼ inches long, about one-third as wide, yellowish-green when unfolding, turning to dark green above, slightly lighter beneath, yellow in autumn; outline lanceolate, oblong to oval, obscurely crenulate-serrate; apex obtuse, scarcely mucronate; base mostly rounded; leafstalks and leaves downy, especially when young.

Inflorescence.—Early June. From lateral or terminal buds on the old wood, in slender, pendent, greenish racemes scarcely distinguishable among the young leaves; sterile and fertile flowers on different trees or on the same tree and even in the same cluster;

calyx somewhat campanulate, 3-5-cleft; petals 3-5, somewhat wider than the sepals, and inserted with the 3-10 stamens on the calyx: pistil in sterile flowers abortive or wanting, conspicuous in the fertile flowers. Parts of the flower more or less pubescent, arachnoid-pubescent within, near the base.

Fruit.—Pods dull red, 1-1½ feet long, flat, pendent, and often twisted, containing several flat brown seeds.

Horticultural Value.—Hardy throughout New England, grows in any well-drained soil, but prefers a deep, rich loam; transplants readily, grows rapidly, is long-lived, free from disease, and makes a picturesque object in ornamental plantations, but is objectionable in public places and highly finished grounds on account of the stiff spines, which are a source of danger to pedestrians, and also on account of the long strap-shaped pods, which litter the ground. There is a thornless form which is better adapted than the type for ornamental purposes. The type is sometimes offered in nurseries at a low price by the quantity. Propagated from seed.

Plate LXVI. — Gleditsia triacanthos.

1. Winter buds.
2. Winter buds with thorns.
3. Flowering branch.
4. Sterile flower, enlarged.
5. Flowering branch, flowers mostly fertile.
6. Fertile flower, enlarged.
7. Fruiting branch.
8. Leaf partially twice pinnate.

[Pg 131]

Robinia Pseudacacia, L.

Locust.

Habitat and Range. — In its native habitat growing upon mountain slopes, along the borders of forests, in rich soils.

Naturalized from Nova Scotia to Ontario.

Maine, — thoroughly at home, forming wooded banks along streams; New Hampshire, — abundant enough to be reckoned among the valuable timber trees; Vermont, — escaped from cultivation in many places; Massachusetts, Rhode Island, and Connecticut, — common in patches and thickets and along the roadsides and fences.

Native from southern Pennsylvania along the mountains to Georgia; west to Iowa and southward.

Habit. — Mostly a small tree, 20-35 feet high, under favorable conditions reaching a height of 50-75 feet; trunk diameter 8 inches to 2 ½ feet; lower branches thrown out horizontally or at a broad angle, forming a few-branched, spreading top, clothed with a tender green, delicate, tremulous foliage, and distinguished in early June by loose, pendulous clusters of white fragrant flowers.

Bark.—Bark of trunk dark, rough and seamy even in young trees, and armed with stout prickles which disappear as the tree matures; in old trees coarsely, deeply, and firmly ridged, not flaky; larger branches a dull brown, rough; branchlets grayish-brown, armed with prickles; season's shoots green, more or less rough-dotted, thin, and often striped.

Winter Buds and Leaves.—Winter buds minute, partially sunken within the leaf-scar. Leaves pinnately compound, alternate; petiole swollen at the base, covering bud of the next season; often with spines in the place of stipules; leaflets 7-21, opposite or scattered, ¾-1¼ inches long, about half as wide, light green; outline ovate or oval-oblong; apex round or obtuse, tipped with a minute point; base truncate, rounded, obtuse or acutish; distinctly short-stalked; stipellate at first. [Pg 132]

Inflorescence.—Late May or early June. Showy and abundant, in loose, pendent, axillary racemes; calyx short, bell-shaped, 5-cleft, the two upper segments mostly coherent; corolla shaped like a pea blossom, the upper petal large, side petals obtuse and separate; style and stigma simple.

Fruit.—A smooth, dark brown, flat pod, about 3 inches long, containing several small brown flattish seeds, remaining on the tree throughout the winter.

Horticultural Value.—Hardy throughout New England in all dry, sunny situations, of rapid growth, spreading by underground stems, ordinarily short-lived and subject to serious injury by the attacks of borers. Occasionally procurable in large quantities at a low rate. In Europe there are many horticultural forms, a few of which are occasionally offered in American nurseries. The type is propagated from seed, the forms by grafting.

Plate LXVII.—Robinia Pseudacacia.

1. Winter buds.
2. Flowering branch.
3. Flower with corolla removed.
4. Fruiting branch.

Robinia viscosa, Vent.

Clammy Locust.

This tree appears to be sparingly established in southern Canada and at many points throughout New England.

Common in cultivation and occasionally established through the middle states; native from Virginia along the mountains of North Carolina, South Carolina, and Georgia.

Easily distinguished from *R. Pseudacacia* by its smaller size, glandular, viscid branchlets, later period of blossoming, and by its more compact, usually upright, scarcely fragrant, rose-colored flower-clusters. [Pg 133]

SIMARUBACEÆ. AILANTHUS FAMILY.

Ailanthus glandulosus, Desf.

Ailanthus. Tree-of-heaven. Chinese Sumac.

Sparsely and locally naturalized in southern Ontario, New England, and southward.

A native of China; first introduced into the United States on an extensive scale in 1820 at Flushing, Long Island; afterwards disseminated by nursery plants and by seed distributed from the Agricultural Department at Washington. Its rapid growth, ability to withstand considerable variations in temperature, and its dark luxuriant foliage made it a great favorite for shade and ornament. It was planted extensively in Philadelphia and New York, and generally throughout the eastern sections of the country. When these trees began to fill the ground with suckers and the vile-scented sterile flowers poisoned the balmy air of June and the water in the cisterns, occasioning many distressing cases of nausea, a reaction set in and hundreds of trees were cut down. The female trees, against the blossoms of which no such objection lay, were allowed to grow, and have often attained a height of 50-75 feet, with a trunk diameter of 3-5 feet. The fruit is very beautiful, consisting of profuse clusters of delicate pinkish or greenish keys.

The tree is easily distinguished by its ill-scented compound leaves, often 2-3 feet long, by the numerous leaflets, sometimes exceeding 40, each ovate, or ovate-lanceolate, with one or two teeth near the base, by its vigorous growth from suckers, and in winter by the coarse, blunt shoots and conspicuous, heart-shaped leaf-scars. [Pg 134]

ANACARDIACEÆ. SUMAC FAMILY.

Rhus typhina, L.

Rhus hirta, Sudw.

Staghorn Sumac.

Habitat and Range. — In widely varying soils and localities; river banks, rocky slopes to an altitude of 2000 feet, cellar-holes and waste places generally, often forming copses.

From Nova Scotia to Lake Huron.

Common throughout New England.

South to Georgia; west to Minnesota and Missouri.

Habit. — A shrub, or small tree, rarely exceeding 25 feet in height; trunk 8-10 inches in diameter; branches straggling, thickish, mostly crooked when old; branchlets forked, straight, often killed at the tips several inches by the frost; head very open, irregular, characterized by its velvety shoots, ample, elegant foliage, turning in early autumn to rich yellows and reds, and by its beautiful, soft-looking crimson cones.

Bark. — Bark of trunk light brown, mottled with gray, becoming dark brownish-gray and more or less rough-scaly in old trees; the season's shoots densely covered with velvety hairs, like the young horns of deer (giving rise to the common name), the pubescence disappearing after two or three years; the extremities dotted with minute orange spots which enlarge laterally in successive seasons, giving a roughish feeling to the branches.

Winter Buds and Leaves. — Buds roundish, obtuse, densely covered with tawny wool, sunk within a large leaf-scar. Leaves pinnately compound, 1-2 feet long; stalk hairy, reddish above, enlarged at base covering the axillary bud; leaflets 11-31, mostly in opposite pairs, the middle pair longest, nearly sessile except the odd one, 2-4 inches long; dark green above, light and often downy beneath; outline narrow to broad-oblong or broad-lanceolate, usually serrate, [Pg 135] rarely laciniate, long-pointed, slightly heart-shaped or rounded at base; stipules none.

Inflorescence. — June to July. Flowers in dense terminal, thyrsoid panicles, often a foot in length and 5-6 inches wide; sterile and fertile mostly on separate trees, but sterile, fertile, and perfect occasionally on the same tree; calyx small, the 5 hairy, ovate-lanceolate sepals united at the base and, in sterile flowers, about half the length of the usually recurved petals; stamens 5, somewhat exserted; ovary abortive, smooth; in the fertile flowers the sepals are nearly as long as the upright petals; stamens short; ovary pubescent, 1-celled, with 3 short styles and 3 spreading stigmas.

Fruit.—In compound terminal panicles, 6-10 or 12 inches long, made up of small, dryish, smooth-stoned drupes densely covered with acid, crimson hairs, persistent till spring.

Horticultural Value.—Hardy throughout New England. Grows in any well-drained soil, but prefers a deep, rich loam. The vigorous growth, bold, handsome foliage, and freedom from disease make it desirable for landscape plantations. It spreads rapidly from suckers, a single plant becoming in a few years the center of a broad-spreading group. Seldom obtainable in nurseries, but collected plants transplant easily.

The cut-leaved form is cultivated in nurseries for the sake of its exceedingly graceful and delicate foliage.

Plate LXVIII.—Rhus typhina.

1. Winter buds.
2. Branch with staminate flowers.
3. Staminate flower.
4. Branch with pistillate flowers.
5. Pistillate flower.
6. Fruit cluster.
7. Fruit.

[Pg 136]

Rhus Vernix, L.

Rhus venenata, DC.

Dogwood. Poison Sumac. Poison Elder.

Habitat and Range.—Low grounds and swamps; occasional on the moist slopes of hills.

Infrequent in Ontario.

Maine,—local and apparently restricted to the southwestern sections; as far north as Chesterville (Franklin county); Vermont,—infrequent; common throughout the other New England states, especially near the seacoast.

South to northern Florida; west to Minnesota and Louisiana.

Habit.—- A handsome shrub or small tree, 5-20 feet high; trunk sometimes 8-10 inches in diameter; broad-topped in the open along the edge of swamps; conspicuous in autumn by its richly colored foliage and diffusely panicled, pale, yellowish-white fruit.

Bark.—Trunk and branches mottled gray, roughish with round spots; branchlets light brown; season's shoots reddish at first, turning later to gray, thickly beset with rough yellowish warts; leaf-scars prominent, triangular.

Buds and Leaves.—Buds small, roundish. Leaves pinnately compound, alternate; rachis abruptly widened at base; leaflets 5-13, opposite, short-stalked except the odd one, 2-3 inches long, 1-2 inches wide, smooth, light green and mostly glossy when young, becoming dark green and often dull, obovate to oval or ovate; entire, often wavy-margined; apex acute, acuminate, or obtuse; base mostly obtuse or rounded; veins prominent, often red; stipules none.

Inflorescence.—Early in July. Near the tips of the branches, in loose, axillary clusters of small greenish flowers; sterile, fertile, and perfect flowers on the same tree, or occasionally sterile and fertile on separate trees; calyx deeply 5-parted, divisions ovate, acute; petals 5, oblong; stamens 5, exserted in the sterile flowers; ovary globose, styles 3. [Pg 137]

Fruit.—Drupes about as large as peas, smooth, more or less glossy, whitish; stone ridged; strongly resembling the fruit of *R. Toxicodendron* (poison ivy).

Horticultural Value.—No large shrub or small tree, so attractive as this, does so well in wet ground; it grows also in any good soil, but it is seldom advisable to use it, on account of its noxious qualities. It can be obtained only from collectors of native plants.

Note.—This sumac has the reputation of being the most poisonous of New England plants. The treacherous beauty of its autumn leaves is a source of grief to collectors. Many are seriously affected, without actual contact, by the exhalation of vapor from the leaves, by grains of pollen floating in the air, and even by the smoke of the burning wood.

It is easily distinguished from the other sumacs. The leaflets are not toothed like those of *R. typhina* (staghorn sumac) and *R. glabra* (smooth sumac); it is not pubescent like *R. typhina* and *R. copallina* (dwarf sumac); the rachis of the compound leaf is not wing-margined as in *R. copallina*; the panicles of flower and fruit are not upright and compact, but drooping and spreading; the fruit is not red-dotted with dense crimson hairs, but is smooth and whitish. Unlike the other sumacs, it grows for the most part in lowlands and swamps.

In the vicinity of Southington, southern Connecticut, *Rhus copallina* is occasionally found with a trunk 5 or 6 inches in diameter (C. H. Bissell).

Plate LXIX. — Rhus Vernix.

1. Winter buds.
2. Branch with sterile flowers.
3. Sterile flower.
4. Branch with fertile flowers.
5. Fertile flower.
6. Fruiting branch.

[Pg 138]

AQUIFOLIACEÆ. HOLLY FAMILY.

Ilex opaca, Ait.

Holly. American Holly.

Habitat and Range. — Generally found in somewhat sheltered situations in sandy loam or in low, moist soil in the vicinity of water.

Maine, — reported on the authority of Gray's *Manual*, sixth edition, in various botanical works, but no station is known; New Hampshire and Vermont, — no station reported; Massachusetts, — occasional from Quincy southward upon the mainland and the island of Naushon; rare in the peat swamps of Nantucket; Rhode Island, — common in South Kingston and Little Compton and sparingly found upon Prudence and Conanicut islands in Narragansett bay; Connecticut, — mostly restricted to the southwestern sections.

Southward to Florida; westward to Missouri and the bottomlands of eastern Texas.

Habit. — A shrub or small tree, exceptionally reaching a height of 30 feet, with a trunk diameter of 15-18 inches, but attaining larger proportions south and west; head conical or dome-shaped, compact; branches irregular, mostly horizontal, clothed with a spiny evergreen foliage. The fertile trees are readily distinguished through late fall and early winter by the conspicuous red berries.

Bark. — Bark of trunk thick, smooth on young trees, roughish, dotted on old, of a nearly uniform ash-gray on trunk and branches; the young shoots more or less downy, bright greenish-yellow, becoming smooth and grayish at the end of the season.

Winter Buds and Leaves. — Buds short, roundish, generally obtuse, scales minutely ciliate. Leaves evergreen, simple, alternate, 2-4 inches long, 1½-3 inches wide, flat when compared with those of the European holly, thickish, smooth on both sides, yellowish-green, scarcely glossy on the upper surface, paler beneath, elliptical, oval or oval-oblong; apex [Pg 139] acutish, spine-tipped; base acutish or obtuse; margin wavy and concave between the large spiny teeth, sometimes with one or two teeth or entire; midrib prominent beneath; leafstalks short, grooved; stipules minute, awl-shaped, becoming blackish, persistent.

Inflorescence. — Flowers in June along the base of the season's shoots; sterile and fertile flowers usually on separate trees, — the sterile in loose, few-flowered clusters, the fertile mostly solitary; peduncles and pedicels slender, bracted midway; calyx persistent, with 4 pointed, ciliate teeth; corolla white, monopetalous, with 4 roundish, oblong divisions; stamens 4, alternating with and shorter than the lobes of the corolla in the fertile flowers, but longer in the sterile; ovary green, nearly cylindrical, surmounted by the sessile, 4-lobed stigma. Parts of the flower sometimes in fives or sixes.

Fruit. — A dull red, berry-like drupe, with 4 nutlets, ribbed or grooved on the convex back, ripening late, and persistent into winter. A yellow-fruited form reported at New Bedford, Mass. (*Rhodora*, III, 58).

Horticultural Value. — Hardy in southern New England; though preferring moist, gravelly loam, it does fairly well in dry soil; of slow growth; useful to form low plantation in shade and to enrich the undergrowth of woods; occasionally sold by collectors but rare in nurseries; nursery plants must be frequently transplanted to be moved successfully; only a small percentage of ordinary collected plants live. The seed seldom germinates in less than two years.

Notes. — The cultivated European holly, which the American tree closely resembles, may be distinguished by its deeper green, glossier, and more wave-margined leaves and the deeper red of its berries.

"There are several fine specimens of the *Ilex opaca* on the farm of Col. Minot Thayer in Braintree, Mass., which are about a foot in diameter a yard above the ground and 25 feet in height. They have maintained their present dimensions for more than fifty years." — D. T. Browne's *Trees of North America*, published in 1846.

This estate is now owned by Mr. Thomas A. Watson. [Pg 140] Several of these trees have been cut down, but one of them is still standing and of substantially the dimensions given above. It must have reached the limit of growth a hundred years ago and now shows very evident signs of decrepitude. This may be due, however, to the loss of a square foot or more of bark from the trunk.

368

Plate LXX. — Ilex opaca.

1. Branch with staminate flowers.
2. Staminate flower.
3. Pistillate flower.
4. Fruiting branch.

ACERACEÆ. MAPLE FAMILY.

Acer rubrum, L.

Red Maple. Swamp Maple. Soft Maple. White Maple.

Habitat and Range. — Borders of streams, low lands, wet forests, swamps, rocky hillsides.

Nova Scotia to the Lake of the Woods.

Common throughout New England from the sea to an altitude of 3000 feet on Katahdin.

South to southern Florida; west to Dakota, Nebraska, and Texas.

Habit. — A medium-sized tree, 40-50 feet high, rising occasionally in swamps to a height of 60-75 feet; trunk 2-4 feet in diameter, throwing out limbs at varying angles a few feet from the ground; branches and branchlets slender, forming a bushy spray, the tips having a slightly upward tendency; head compact, in young trees usually rounded and symmetrical, widest just above the point of furcation. In the first warm days of spring there shimmers amid the naked branches a faint glow of red, which at length becomes embodied in the abundant scarlet, crimson, or yellow of the long flowering stems; succeeded later by the brilliant fruit, which is outlined against the sober green of the foliage till it pales and falls in [Pg 141] June. The colors of the autumn leaves vie in splendor with those of the sugar maple.

Bark. — In young trees smooth and light gray, becoming very dark and ridgy in large trunks, the surface separating into scales, and in very old trees hanging in long flakes; young shoots often bright red in autumn, conspicuously marked with oblong white spots.

Winter Buds and Leaves. — Buds aggregated at or near the ends of the preceding year's shoots, about ⅛ inch long; protected by dark reddish scales; inner scales lengthening with the growth of the shoot. Leaves simple, opposite, 3-4 inches long, green and smooth above, lighter and more or less pubescent beneath, especially along the veins; turning crimson or scarlet in early autumn; ovate, 3-5-lobed, the middle lobe generally the longest, the lower pair (when 5 lobes are present) the smallest; unequally sharp-toothed, with broad, acute sinuses; apex acute; base heart-shaped, truncate, or obtuse; leafstalk 1-3 inches long. The leaves of the red maple vary greatly in size, outline, lobing, and shape of base.

Inflorescence. — April 1-15. Appearing before the leaves in close clusters encircling the shoots of the previous year, varying in color from dull red or pale yellow to scarlet; the sterile and fertile flowers mostly in separate clusters, sometimes on the same tree, but more frequently on different trees; calyx lobes oblong and obtuse; petals linear-oblong; pedicels short, stamens 5-8, much longer than the petals in the sterile and about the same length in the fertile flowers; the smooth ovary surmounted by a style separating into two much-projecting stigmatic lobes.

Fruit. — Fruit ripe in June, hanging on long stems, varying from brown to crimson; keys about an inch in length, at first convergent, at maturity more or less divergent.

Horticultural Value. — Hardy throughout New England; found in a wider range of soils than any other species of the genus, but seeming to prefer a gravelly or peaty loam in positions where its roots can reach a constant supply of moisture. It is more variable than any other of the native maples and consequently is not so good a tree for streets, where a [Pg 142] symmetrical outline and uniform habit are required. It is transplanted readily, but recovers its vigor more slowly than does the sugar or silver maple and is usually of slower growth. Its variable habit makes it an exceedingly interesting tree in the landscape.

Plate LXXI. — Acer rubrum.

1. Leaf-buds.
2. Flower-buds.
3. Branch with sterile flowers.
4. Sterile flower.
5. Branch with sterile and fertile flowers.
6. Fertile flower.
7. Fruiting branch.
8. Variant leaves.

Acer saccharinum, L.

Acer dasycarpum, Ehrh.

Silver Maple. Soft Maple. White Maple. River Maple.

Habitat and Range. — Along streams, in rich intervale lands, and in moist, deep-soiled forests, but not in swamps.

Infrequent from New Brunswick to Ottawa, abundant from Ottawa throughout Ontario.

Occasional throughout the New England states; most common and best developed upon the banks of rivers and lakes at low altitudes.

South to the Gulf states; west to Dakota, Nebraska, Kansas, and Indian territory; attaining its maximum size in the basins of the Ohio and its tributaries; rare towards the seacoast throughout the whole range.

Habit. — A handsome tree, 50-60 feet in height; trunk 2-5 feet in diameter, separating a few feet from the ground into several large, slightly diverging branches. These, naked for some distance, repeatedly subdivide at wider angles, forming a very wide head, much broader near the top. The ultimate branches are long and slender, often forming on the lower [Pg 143] limbs a pendulous fringe sometimes reaching to the ground. Distinguished in winter by its characteristic graceful outlines, and by its flower-buds conspicuously scattered along the tips of the branchlets; in summer by the silvery-white under-surface of its deeply cut leaves. It is among the first of the New England trees to blossom, preceding the red maple by one to three weeks.

Bark. — Bark of trunk smooth and gray in young trees, becoming with age rougher and darker, more or less ridged, separating into thin, loose scales; young shoots chestnut-colored in autumn, smooth, polished, profusely marked with light dots.

Winter Buds and Leaves. — Flower-buds clustered near the ends of the branchlets, conspicuous in winter; scales imbricated, convex, polished, reddish, with ciliate margins; leaf-buds more slender, about ⅛ inch long, with similar scales, the inner lengthening, falling as the leaf expands. Leaves simple, opposite, 3-5 inches long, of varying width, light green above, silvery-white beneath, turning yellow in autumn; lobes 3, or more usually 5, deeply cut, sharp-toothed, sharp-pointed, more or less sublobed; sinuses deep, narrow, with concave sides; base sub-heart-shaped or truncate; stems long.

Inflorescence.—March to April. Much preceding the leaves; from short branchlets of the previous year, in simple, crowded umbels; flowers rarely perfect, the sterile and fertile sometimes on the same tree and sometimes on different trees, generally in separate clusters, yellowish-green or sometimes pinkish; calyx 5-notched, wholly included in bud-scales; petals none; sterile flowers long, stamens 3-7 much exserted, filaments slender, ovary abortive or none: fertile flowers broad, stamens about the length of calyx-tube, ovary woolly, with two styles scarcely united at the base.

Fruit.—Fruit ripens in June, earliest of the New England maples. Keys large, woolly when young, at length smooth, widely divergent, scythe-shaped or straight, yellowish-green, one key often aborted.

Horticultural Value.—Hardy in cultivation throughout New England. The grace of its branches, the beauty of its foliage, [Pg 144] and its rapid growth make it a favorite ornamental tree. It attains its finest development when planted by the margin of pond or stream where its roots can reach water, but it grows well in any good soil. Easily transplanted, and more readily obtainable at a low price than any other tree in general use for street or ornamental purposes. The branches are easily broken by wind and ice, and the roots fill the ground for a long distance and exhaust its fertility.

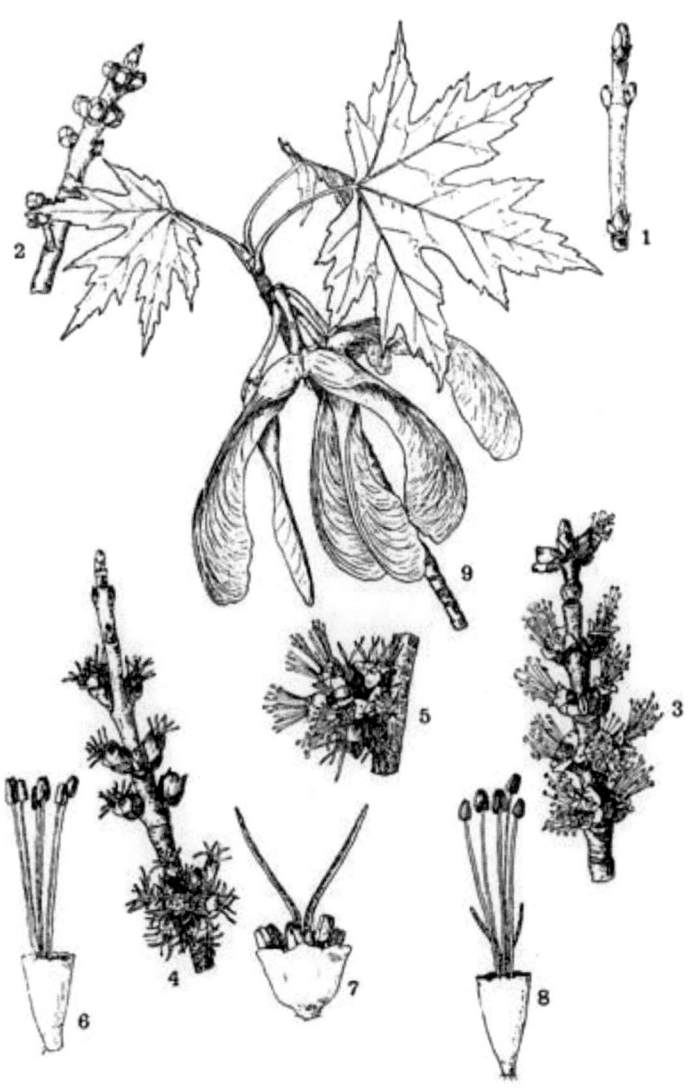

Plate LXXII. — Acer saccharinum.

1. Leaf-buds.
2. Flower-buds.
3. Branch with sterile flowers.
4. Branch with fertile flowers.
5. Branch with sterile and fertile flowers.
6. Sterile flower.
7. Fertile flower.
8. Perfect flower.
9. Fruiting branch.

Acer Saccharum, Marsh.

Acer saccharinum, Wang. Acer barbatum, Michx.

Rock Maple. Sugar Maple. Hard Maple. Sugar Tree.

Habitat and Range. — Rich woods and cool, rocky slopes.

Nova Scotia and Newfoundland, westward to Lake of the Woods.

New England, — abundant, distributed throughout the woods, often forming in the northern portions extensive upland forests; attaining great size in the mountainous portions of New Hampshire and Vermont, and in the Connecticut river valley; less frequent toward the seacoast.

South to the Gulf states; west to Minnesota, Nebraska, Kansas, and Texas.

Habit. — A noble tree, 50-90 feet in height; trunk 2-5 feet in diameter, stout, erect, throwing out its primary branches [Pg 145] at acute angles; secondary branches straight, slender, nearly horizontal or declining at the base, leaving the stem higher up at sharper and sharper angles, repeatedly subdividing, forming a dense and rather stiff spray of nearly uniform length; head symmetrical, varying greatly in shape; in young trees often narrowly cylindrical, becoming pyramidal or broadly egg-shaped with age; clothed with dense masses of foliage, purple-tinged in spring, light green in summer, and gorgeous beyond all other trees of the forest, with the possible exception of the red maple, in its autumnal oranges, yellows, and reds.

Bark. — Bark of trunk and principal branches gray, very smooth, close and firm in young trees, in old trees becoming deeply furrowed, often cleaving up at one edge in long, thick, irregular plates; season's shoots at length of a shining reddish-brown, smooth, numerously pale-dotted, turning gray the third year.

Winter Buds and Leaves. — Buds sharp-pointed, reddish-brown, minutely pubescent, terminal ¼ inch long, lateral ⅛ inch, appressed, the inner scales lengthening with the growth of the shoot. Leaves simple, opposite, 3-5 inches long, with a somewhat greater breadth, purplish and more or less pubescent when opening, at maturity dark green above, paler, with or without pubescence beneath, changing to brilliant reds and yellows in autumn; lobes sometimes

3, usually 5, acuminate, sparingly sinuate-toothed, with shallow, rounded sinuses; base subcordate, truncate, or wedge-shaped; veins and veinlets conspicuous beneath; leafstalks long, slender.

Inflorescence. — April 1-15. Appearing with the leaves in nearly sessile clusters, from terminal and lateral buds; flowers greenish-yellow, pendent on long thread-like, hairy stems; sterile and fertile on the same or on different trees, usually in separate, but not infrequently in the same cluster; the 5-lobed calyx cylindrical or bell-shaped, hairy; petals none; stamens 6-8, in sterile flowers much longer than the calyx, in fertile scarcely exserted; ovary smooth, abortive in sterile flowers, in fertile surmounted by a single style with two divergent, thread-like, stigmatic lobes. [Pg 146]

Fruit. — Keys usually an inch or more in length, glabrous, wings broad, mostly divergent, falling late in autumn.

Horticultural Value. — Hardy throughout New England. Its long life, noble proportions, beautiful foliage, dense shade, moderately rapid growth, usual freedom from disease or insect disfigurement, and adaptability to almost any soil not saturated with water make it a favorite in cultivation; readily obtainable in nurseries, transplants easily, recovers its vigor quickly, and has a nearly uniform habit of growth.

Note. — Not liable to be taken for any other native maple, but sometimes confounded with the cultivated Norway maple, *Acer platanoides*, from which it is easily distinguished by the milky juice which exudes from the broken petiole of the latter.

The leaves of the Norway maple are thinner, bright green and glabrous beneath, and its keys diverge in a straight line.

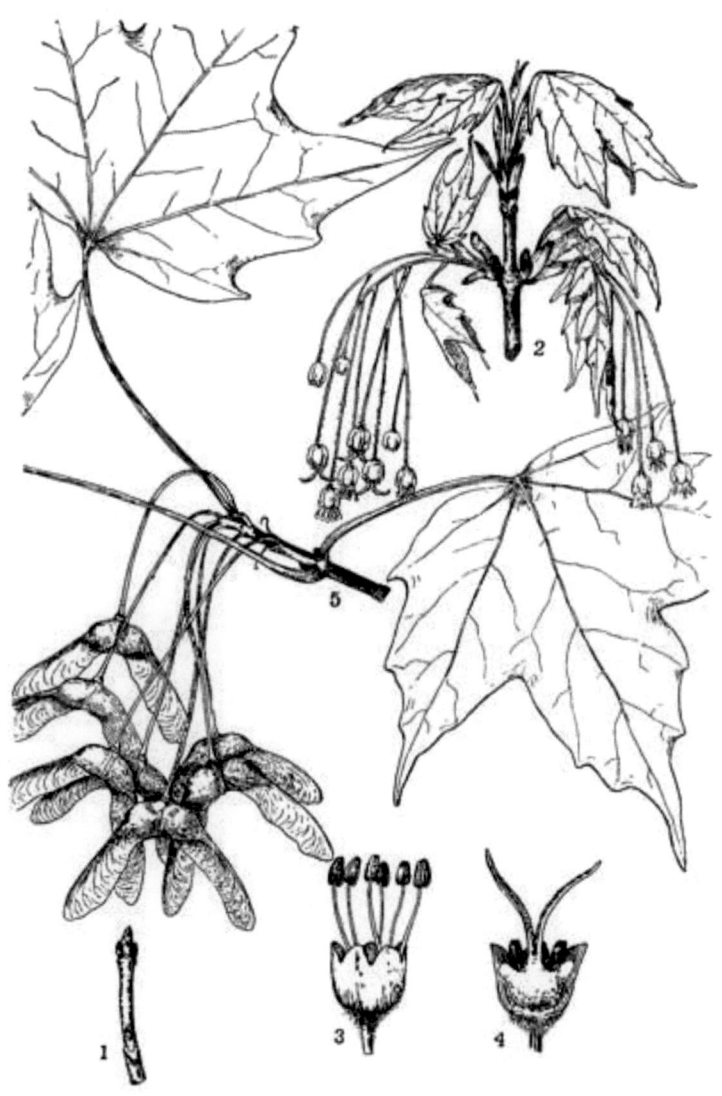

383

Plate LXXIII. — Acer saccharum.

1. Winter buds.
2. Flowering branch.
3. Sterile flower.
4. Fertile flower, part of perianth and stamens removed.
5. Fruiting branch.

Acer saccharum, Marsh., var. nigrum, Britton.

Acer nigrum, Michx. Acer saccharinum, **var.** *nigrum, T. & G. Acer barbatum,* **var.** *nigrum, Sarg.*

Black Maple.

Habitat and Range.—Low, damp ground on which, in New England at least, the sugar maple is rarely if ever seen, or upon moist, rocky slopes.

Apparently a common tree from Ottawa westward throughout Ontario.

The New England specimens, with the exception of those from the Champlain valley, appear to be dubious intermediates between the type and the variety. [Pg 147]

Maine,—the Rangeley lake region; New Hampshire,—occasional near the Connecticut river; Vermont,—frequent in the western part in the Champlain valley, occasional in all other sections, especially in the vicinity of the Connecticut; Massachusetts,—occasional in the Connecticut river valley and westward, doubtfully reported from eastern sections; Rhode Island,—doubtful, resting on the authority of Colonel Olney's list; Connecticut,—doubtfully reported.

South along the Alleghanies to the Gulf states; west to the 95th meridian.

The extreme forms of *nigrum* show well-marked varietal differences; but there are few, if any, constant characters. Further research in the field is necessary to determine the status of these interesting plants.

Habit.—The black maple is somewhat smaller than the sugar maple, the bark is darker and the foliage more sombre. It generally has a symmetrical outline, which it retains to old age.

Leaves.—The fully grown leaves are often larger than those of the type, darker green above, edges sometimes drooping, width equal to or exceeding the length, 5-lobed, margin blunt-toothed, wavy-toothed, or entire, the two lower lobes small, often reduced to a curve in the outline, broad at the base, which is usually heart-shaped; texture firm; the lengthening scales of the opening leaves, the young shoots, the petioles, and the leaves themselves are covered with a downy to a densely woolly pubescence. As the parts mature, the woolliness usually disappears, except along the midrib and principal veins, which become almost glabrous.

Horticultural Value.—Hardy throughout New England, preferring a moist, fertile, gravelly loam; young trees are rather more vigorous than those of the sugar maple, and easily transplanted. Difficult to secure, for it is seldom offered for sale or recognized by nurseries, although occasionally found mixed with the sugar maple in nursery rows.

Plate LXXIV. — Acer Saccharum, var. nigrum.

1. Fruiting branch.

[Pg 148]

Acer spicatum, Lam.

Mountain Maple.

Habitat and Range. — In damp forests, rocky highland woods, along the sides of mountain brooks at altitudes of 500-1000 feet.

From Nova Scotia and Newfoundland to Saskatchewan.

Maine, — common, especially northward in the forests; New Hampshire and Vermont, — common; Massachusetts, — rather common in western and central sections, occasional eastward; Rhode Island, — occasional northward; Connecticut, — occasional in northern and central sections; reported as far south as North Branford (New Haven county).

Along mountain ranges to Georgia.

Habit. — Mostly a shrub, but occasionally attaining a height of 25 feet, with a diameter, near the ground, of 6-8 inches; characterized by a short, straight trunk and slender branches; bright green foliage turning a rich red in autumn, and long-stemmed, erect racemes of delicate flowers, drooping at length beneath the weight of the maturing keys.

Bark. — Bark of trunk thin, smoothish, grayish-brown; primary branches gray; branchlets reddish-brown streaked with green, retaining in the second year traces of pubescence; season's shoots yellowish-green, reddish on the upper side when exposed to the sun, minutely pubescent.

Winter Buds and Leaves. — Buds small, flattish, acute, slightly divergent from the stem. Leaves simple, opposite, 4-5 inches long, two-thirds as wide, pubescent on both sides when unfolding, at length glabrous on the upper surface, 3-lobed above the center, often with two small additional lobes at the base, coarsely or finely serrate, lobes acuminate; base more or less heart-shaped; veining 3-5-nerved, prominent, especially on the lower side, furrowed above; leafstalks long, enlarged at the base.

Inflorescence.—June. Appearing after the expansion of the leaves, in long-stemmed, terminal, more or less panicled, erect [Pg 149] or slightly drooping racemes; flowers small and numerous, both kinds in the same raceme, the fertile near the base; all upon very slender pedicels; lobes of calyx 5, greenish, downy, about half as long as the alternating linear petals; stamens usually 8, in the sterile flower nearly as long as the petals, in the fertile much shorter; pistil rudimentary, hairy in the sterile flower; in the fertile the ovary is surmounted by an erect style with short-lobed stigma.

Fruit.—In long racemes, drooping or pendent; the keys, which are smaller than those of any other American maple, set on hair-like pedicels, and at a wide but not constant angle; at length reddish, with a small cavity upon one side.

Horticultural Value.—Hardy in cultivation throughout New England; prefers moist, well-drained, gravelly loam in partial shade, but grows well in any good soil; easily transplanted, but recovers its vigor rather slowly; foliage free from disease.

Seldom grown in nurseries, but readily obtainable from northern collectors of native plants.

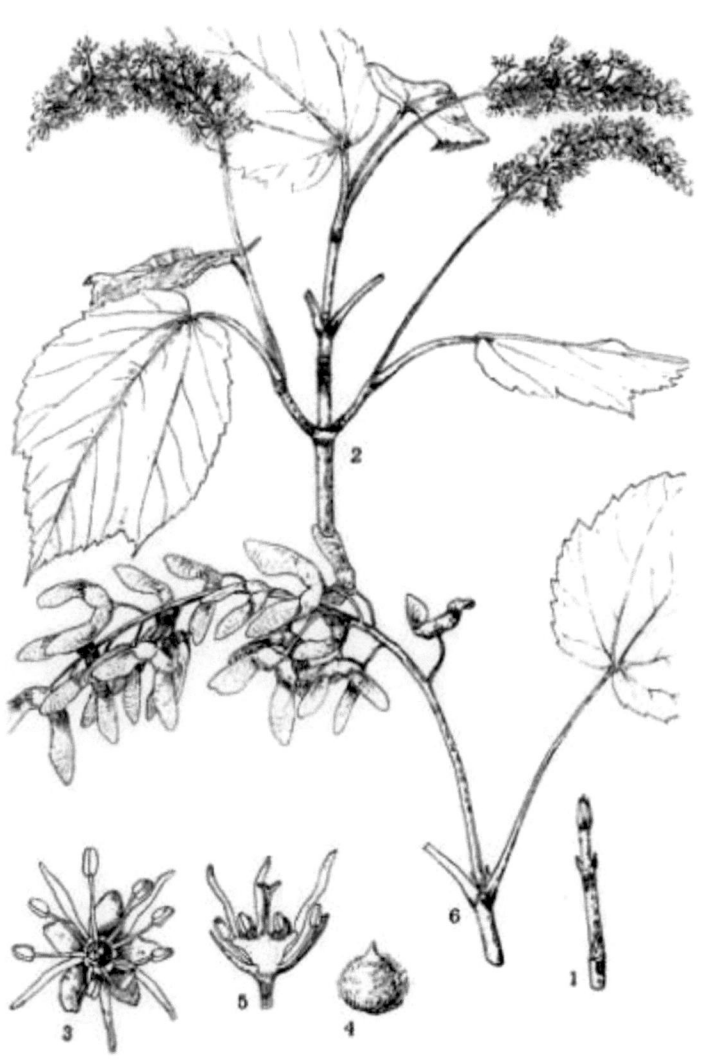

Plate LXXV.—Acer spicatum.

1. Winter buds.
2. Flowering branch.
3. Sterile flower.
4. Abortive ovary in sterile flower.
5. Fertile flower with part of the perianth and stamens removed.
6. Fruiting branch.

Acer Pennsylvanicum, L.

Striped Maple. Moosewood. Whistlewood.

Habitat and Range.—Cool, rocky or sandy woods.

Nova Scotia to Lake Superior.

Maine,—abundant, especially northward in the forests; New Hampshire and Vermont,—common in highland woods; Massachusetts,—common in the western and central sections, rare towards the coast; Rhode Island,—frequent northward; [Pg 150] Connecticut,—frequent, reported as far south as Cheshire (New Haven county).

South on shaded mountain slopes and in deep ravines to Georgia; west to Minnesota.

Habit.—Shrub or small tree, 15-25 feet high, with a diameter at the ground of 5-8 inches; characterized by a slender, beautifully striate trunk and straight branches; by the roseate flush of the opening foliage, deepening later to a yellowish-green; and by the long, graceful, pendent racemes of yellowish flowers, succeeded by the abundant, drooping fruit.

Bark.—Bark of trunk and branches deep reddish-brown or dark green, conspicuously striped longitudinally with pale and blackish bands; roughish with light buff, irregular dots; the younger branches marked with oval leaf-scars and the linear scars of the leaf-scales; the season's shoots smooth, light green, mottled with black.

In spring the bark of the small branches is easily separable, giving rise to the name "whistle wood."

Winter Buds and Leaves. — Terminal bud long, short-stalked, obscurely 4-sided, tapering to a blunt tip; lateral buds small and flat; opening foliage roseate. Leaves simple, opposite; 5-6 inches long and nearly as broad; the upper leaves much narrower; when fully grown light green above, paler beneath, finally nearly glabrous, yellow in autumn, divided above the center into three deep acuminate lobes, finely, sharply, and usually doubly serrate; base heart-shaped, truncate, or rounded; leafstalks 1-3 inches long, grooved, the enlarged base including the leaf-buds of the next season.

Inflorescence. — In simple, drooping racemes, often 5-6 inches long, appearing after the leaves in late May or early June; the sterile and fertile flowers mostly in separate racemes on the same tree; the bell-shaped flowers on slender pedicels; petals and sepals greenish-yellow; sepals narrowly oblong, somewhat shorter than the obovate petals; stamens usually 8, shorter than the petals in the sterile flower, rudimentary in the fertile, the pistil abortive or none in the sterile flower, in the fertile terminating in a recurved stigma. [Pg 151]

Fruit. — In long, drooping racemes of pale green keys, set at a wide but not uniform angle; distinguished from the other maples, except *A. spicatum*, by a small cavity in the side of each key; abundant; ripening in August.

Horticultural Value. — Hardy, under favorable conditions, throughout New England. Prefers a rich, moist soil near water, in shade; but grows well in almost any soil when once established, many young plants failing to start into vigorous growth. Occasionally grown by nurserymen, but more readily obtainable from northern collectors of native plants.

Plate LXXVI. — Acer Pennsylvanicum.

1. Winter buds.
2. Flowering branch.
3. Sterile flower.
4. Fertile flower with part of the perianth removed.
5. Fruiting branch.

Acer Negundo, L.

Negundo aceroides, Moench. Negundo Negundo, Karst.

Box Elder. Ash-leaved Maple.

Habitat and Range. — In deep, moist soil; river valleys and borders of swamps.

Infrequent from eastern Ontario to Lake of the Woods; abundant from Manitoba westward to the Rocky mountains south of 55° north latitude.

Maine, — along the St. John and its tributaries, especially in the French villages, the commonest roadside tree, brought in from the wild state according to the people there; thoroughly established young trees, originating from planted specimens, in various parts of the state; New Hampshire, — occasional along the Connecticut, abundant at Walpole; extending northward as far as South Charlestown (W. F. Flint *in lit.*); Vermont, — shores of the Winooski river and of Lake Champlain; Connecticut, — banks of the Housatonic river at New Milford, Cornwall Bridge, and Lime Rock station. [Pg 152]

South to Florida; west to the Rocky and Wahsatch mountains, reaching its greatest size in the river bottoms of the Ohio and its tributaries.

Habit. — A small but handsome tree, 30-40 feet high, with a diameter of 1-2 feet. Trunk separating at a small height, occasionally a foot or two from the ground, into several wide-spreading branches, forming a broad, roundish, open head, characterized by lively green branchlets and foliage, delicate flowers and abundant, long, loose racemes of yellowish-green keys hanging till late autumn, the stems clinging throughout the winter.

Bark. — Bark of trunk when young, smooth, yellowish-green, in old trees becoming grayish-brown and ridgy; smaller branchlets greenish-yellow; season's shoots pale green or sometimes reddish-purple, smooth and shining or sometimes glaucous.

Winter Buds and Leaves. — Buds small, ovate, enclosed in two dull-red, minutely pubescent scales. Leaves pinnately compound, opposite; leaflets usually 3, sometimes 5 or 7, 2-4 inches long, 1½-2½ inches broad, light green above, paler beneath and woolly when opening, slightly pubescent at maturity, ovate or oval, irregularly and remotely coarse-toothed mostly above the middle, 3-lobed or nearly entire; apex acute; base extremely variable; veins prominent; petioles 2-3 inches long, enlarging at the base, leaving, when they fall, conspicuous leaf-scars which unite at an angle midway between the winter buds.

Inflorescence.—April 1-15. Flowers appearing at the ends of the preceding year's shoots as the leaf-buds begin to open, small, greenish-yellow; sterile and fertile on separate trees,—the sterile in clusters, on long, hairy, drooping, thread-like stems; the calyx hairy, 5-lobed, with about 5 hairy-stemmed, much-projecting linear anthers; pistil none: the fertile in delicate, pendent racemes, scarcely distinguishable at a distance from the foliage; ovary pubescent, rising out of the calyx; styles long, divergent; stamens none.

Fruit.—Loose, pendent, greenish-yellow racemes, 6-8 inches long, the slender-pediceled keys joined at a wide angle, broadest [Pg 153] and often somewhat wavy near the extremity, dropping in late autumn from the reddish stems, which hang on till spring.

Horticultural Value.—Hardy throughout New England; flourishes best in moist soil near running water or on rocky slopes, but accommodates itself to almost any situation; easily transplanted. Plants of the same age are apt to vary so much in size and habit as to make them unsuitable for street planting.

An attractive tree when young, especially when laden with fruit in the fall. There are several horticultural varieties with colored foliage, some of which are occasionally offered in nurseries. A western form, having the new growth covered with a glaucous bloom, is said to be longer-lived and more healthy than the type.

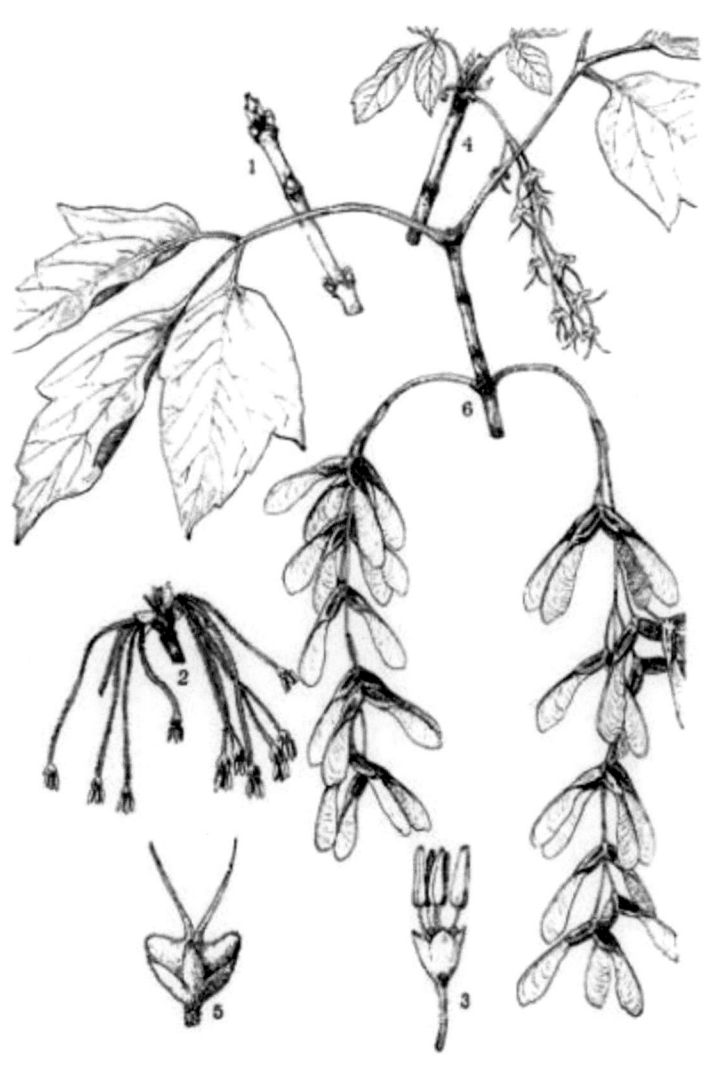

Plate LXXVII. — Acer Negundo.

1. Winter buds.
2. Branch with sterile flowers.
3. Sterile flower.
4. Branch with fertile flowers.
5. Fertile flower.
6. Fruiting branch.

TILIACEÆ. LINDEN FAMILY.

Tilia Americana, L.

Basswood. Linden. Lime. Whitewood.

Habitat and Range. — In rich woods and loamy soils.

Southern Canada from New Brunswick to Lake Winnipeg.

Throughout New England, frequent from the seacoast to altitudes of 1000 feet; rare from 1000 to 2000 feet.

South along the mountains to Georgia; west to Kansas, Nebraska, and Texas.

Habit. — A large tree, 5O-75 feet high, rising in the upper valley of the Connecticut river to the height of 100 feet; trunk 2-4 feet in diameter, erect, diminishing but slightly to [Pg 154] the branching point; head, in favorable situations, broadly ovate to oval, rather compact, symmetrical; branches mostly straight, striking out in different trees at varying angles; the numerous secondary branches mostly horizontal, slender, often drooping at the extremities, repeatedly subdividing, forming a dense spray set at broad angles. Foliage very abundant, green when fully grown, almost impervious to sunlight; the small creamy flowers in numerous clusters; the pale, odd-shaped bracts and pea-like fruit conspicuous among the leaves till late autumn.

Bark. — Dark gray, very thick, smooth in young trees, later becoming broadly and firmly ridged; in old trees irregularly furrowed; branches, especially upon the upper side, dark brown and blackish; the season's shoots yellowish-green to reddish-brown, and numerously rough-dotted. The inner bark is fibrous and tough.

Winter Buds and Leaves. — Leaf-buds small, conical, brownish red, contrasting strongly with the dark stems. Leaves simple, alternate, 4-5 inches long, three-fourths as wide, green and smooth on both sides, thickish, paler beneath, broad-ovate, one-sided, serrate, the point often incurved; apex acuminate or acute; base heart-shaped to truncate; midrib and veins conspicuous on the under surface with minute, reddish tufts of down at the angles; stems smooth, 1-1½ inches long; stipules soon falling.

Inflorescence. — Late June or early July. In loose, slightly fragrant, drooping cymes, the peduncle attached about half its length to a narrowly oblong, yellowish bract, obtuse at both ends, free at the top, and tapering slightly at the base, pedicels slender; calyx of 5 colored sepals united toward the base; corolla of 5 petals alternate with the sepals, often obscurely toothed at the apex; 5 petal-like scales in front of the petals and nearly as long; calyx, petals, and scales yellowish-white; stamens indefinite, mostly in clusters inserted with the scales; anthers 2-celled, ovary 5-celled; style 1; stigma 5-toothed.

Fruit. — About the size of a pea, woody, globose, pale green, 1-celled by abortion: 1-2 seeds. [Pg 155]

Horticultural Value. — Useful as an ornamental or street tree; hardy throughout New England, easily transplanted, and grows rapidly in almost any well-drained soil; comes into leaf late and drops its foliage in early fall. The European species are more common in nurseries. They are, however, seriously affected by wood borers, while the native tree has few disfiguring insect enemies. Usually propagated from the seed. A horticultural form with weeping branches is sometimes cultivated.

Note. — There is so close a resemblance between the lindens that it is difficult to distinguish the American species from each other, or from their European relatives.

American species sometimes found in cultivation:

Tilia pubescens, Ait., is distinguished from *Americana* by its smaller, thinner leaves and densely pubescent shoots.

Tilia heterophylla, Vent., is easily recognized by the pale or silver white under-surface of the leaves.

There are several European species more or less common in cultivation, indiscriminately known in nurseries as *Tilia Europæa*. They are all easily distinguished from the American species by the absence of petal-like scales.

406

Plate LXXVIII. — Tilia Americana.

1. Winter buds.
2. Flowering branch.
3. Flower enlarged.
4. Pistil with cluster of stamens, petaloid scale, petal, and sepal.
5. Fruiting branch.

[Pg 156]

CORNACEÆ. DOGWOOD FAMILY.

Cornus florida, L.

Flowering Dogwood. Boxwood.

Habitat and Range. — Woodlands, rocky hillsides, moist, gravelly ridges.

Provinces of Quebec and Ontario.

Maine, — Fayette Ridge, Kennebec county; New Hampshire, — along the Atlantic coast and very near the Connecticut river, rarely farther north than its junction with the West river; Vermont, — southern and southwestern sections, rare; Massachusetts, — occasional throughout the state, common in the Connecticut river valley, frequent eastward; Rhode Island and Connecticut, — common.

South to Florida; west to Minnesota and Texas.

Habit. — A small tree, 15-30 feet high, with a trunk diameter of 6-10 inches. The spreading branches form an open, roundish head, the young twigs curving upwards at their extremities. In spring, when decked with its abundant, showy white blossoms, it is the fairest of the minor trees of the forest; in autumn, scarcely less beautiful in the rich reds of its foliage and fruit.

Bark. — Bark of trunk in old trees blackish, broken-ridged, rough, often separating into small, firm, 4-angled or roundish plates;

branches grayish, streaked with white lines; season's twigs purplish-green, downy; taste bitter.

Winter Buds and Leaves.—Terminal leaf-buds narrowly conical, acute; flower-buds spherical or vertically flattened, grayish. Leaves simple, opposite, 3-5 inches long, two-thirds as wide, dark green above, whitish beneath, turning to reds, purples, and yellows in the autumn, ovate to oval, nearly smooth, with minute appressed pubescence on both surfaces; apex pointed; base acutish; veins distinctly indented above, ribs curving upward and parallel; leafstalk short-grooved.

Inflorescence.—May to June. Appearing with the unfolding leaves in close clusters at the ends of the branches, each [Pg 157] cluster subtended by a very conspicuous 4-leafed involucre (often mistaken for the corolla and constituting all the beauty of the blossom), the leaves of which are white or pinkish, 1½ inches long, obovate, curiously notched at the rounded end. The real flowers are insignificant, suggesting the tubular disk flowers of the Compositæ; calyx-tube coherent with the ovary, surmounting it by 4 small teeth; petals greenish-yellow, oblong, reflexed; stamens 4; pistil with capitate style.

Fruit.—Ovoid, scarlet drupes, about ½ inch long, united in clusters, persistent till late autumn or till eaten by the birds.

Horticultural Value.—Hardy in southern and southern-central New England, but liable farther north to be killed outright or as far down as the surface of the snow; not only one of the most attractive small trees on account of its flowers, habit, and foliage, but one of the most useful for shady places or under tall trees. The species, a red-flowering and also a weeping variety are obtainable in leading nurseries. Collected plants can be made to succeed. It is a plant of rather slow growth.

Plate LXXIX.—Cornus florida.

1. Leaf-buds.
2. Flower-buds.
3. Flowering branch.
4. Flower.
5. Fruiting branch.

Cornus alternifolia, L. f.

Dogwood. Green Osier.

Habitat and Range.—Hillsides, open woods and copses, borders of streams and swamps.

Nova Scotia and New Brunswick along the valley of the St. Lawrence river to the western shores of Lake Superior.

Common throughout New England.

South to Georgia and Alabama; west to Minnesota.

[Pg 158]

Habit.—A shrub or small tree, 6-20 feet high, trunk diameter 3-6 inches; head usually widest near the top, flat; branches nearly horizontal with lateral spray, the lively green, dense foliage lying in broad planes.

Bark.—Trunk and larger branches greenish, warty, streaked with gray; season's shoots bright yellowish-green or purplish, oblong-dotted.

Winter Buds and Leaves.—Buds small, acute. Leaves simple, alternate or sometimes opposite, clustered at the ends of the branchlets, 2-4 inches long, dark green on the upper side, paler beneath, with minute appressed pubescence on both sides, ovate to oval, almost entire; apex long-pointed; base acutish or rounded; veins indented above, ribs curving upward and parallel; petiole long, slender, and grooved.

Inflorescence. — June. From shoots of the season, in irregular open cymes; calyx coherent with ovary, surmounting it by 4 minute teeth; corolla white or pale yellow, with the 4 oblong petals at length reflexed: stamens 4, exserted; style short, with capitate stigma.

Fruit. — October. Globular, blue or blue black, on slender, reddish stems.

Horticultural Value. — Hardy throughout New England, adapting itself to a great variety of situations, but preferring a soil that is constantly moist. Nursery or good collected plants are easily transplanted. A disease, similar in its effect to the pear blight, so often disfigures it that it is not desirable for use in important plantations.

413

Plate LXXX.—Cornus alternifolia.

1. Winter buds.
2. Flowering branch.
3. Flower with one petal and two stamens removed, side view.
4. Flower, view from above.
5. Fruiting branch.

[Pg 159]

Nyssa sylvatica, Marsh.

Tupelo. Sour Gum. Pepperidge.

Habitat and Range.—In rich, moist soil, in swamps and on the borders of rivers and ponds.

Ontario.

Maine,—Waterville on the Kennebec, the most northern station yet reported (Dr. Ezekiel Holmes); New Hampshire,—most common in the Merrimac valley, seldom seen north of the White mountains; Vermont,—occasional; Massachusetts, Rhode Island, and Connecticut,—rather common.

South to Florida; west to Michigan, Missouri, and Texas.

Habit.—Tree 20-50 feet high, with a trunk diameter of 1-2 feet, rising in the forest to the height of 60-80 feet; attaining greater dimensions farther south; lower branches horizontal or declining, often touching the ground at their tips, the upper horizontal or slightly rising, angular, repeatedly subdividing; branchlets very numerous, short and stiff, making a flat spray; head extremely variable, unique in picturesqueness of outline; usually broad-spreading, flat-topped or somewhat rounded; often reduced in Nantucket and upon the southern shore of Cape Cod to a shrub or small tree of 10-15 feet in height, forming low, dense, tangled thickets. Foliage very abundant, dark lustrous green, turning early in the fall to a brilliant crimson.

Bark.—Trunk of young trees grayish-white, with irregular and shallow striations, in old trees darker, breaking up into somewhat hexagonal or lozenge-shaped scales; branches smooth and brown; season's shoots reddish-green, with a few minute dots.

Winter Buds and Leaves.—Buds ovoid, ⅛-¼ inch long, obtuse. Leaves simple, irregularly alternate, often apparently whorled when clustered at the ends of the shoots, 2-5 inches long, one-half as wide; at first bright green beneath, dullish-green above, becoming dark glossy green above, paler beneath, obovate or oblanceolate to oval; entire, few or [Pg 160] obscurely toothed, or wavy-margined above the center; apex more or less abruptly acute; base acutish; firm, smooth, finely sub-veined; stem short, flat, grooved, minutely ciliate, at least when young; stipules none.

Inflorescence.—May or early June. Appearing with the leaves in axillary clusters of small greenish flowers, sterile and fertile usually on separate trees, sometimes on the same tree,—sterile flowers in simple or compound clusters; calyx minutely 5-parted, petals 5, small or wanting; stamens 5-12, inserted on the outside of a disk; pistil none: fertile flowers larger, solitary, or several sessile in a bracted cluster; petals 5, small or wanting; calyx minutely 5-toothed.

Fruit.—Drupes 1-several, ovoid, blue black, about ½ inch long, sour: stone striated lengthwise.

Horticultural Value.—Hardy throughout New England; adapts itself readily to most situations but prefers deep soil near water. Seldom offered in nurseries and difficult to transplant unless frequently root-pruned or moved; collected plants do not thrive well; seedlings are raised with little difficulty. Few trees are of greater ornamental value.

Plate LXXXI. — Nyssa sylvatica.

1. Winter buds.
2. Branch with sterile flowers.
3-4. Sterile flowers.
5. Branch with fertile flowers.
6. Fertile flower.
7. Fruiting branch.

EBENACEÆ. EBONY FAMILY.

Diospyros Virginiana, L.

Persimmon.

Habitat and Range. — Rhode Island, — occasional but doubtfully native; Connecticut, — at Lighthouse Point, New Haven, near the East Haven boundary line, there is a grove consisting of about one hundred twenty-five small trees not more than a hundred feet from the water's edge, in sandy soil just [Pg 161] above the beach grass, exposed to the buffeting of fierce winds and the incursions of salt water, which comes up around them during the heavy winter storms. These trees are not in thriving condition; several are dead or dying, and no new plants are springing up to take their places. A cross-section of the trunk of a dead tree, as large as any of those living, shows about fifty annual rings. There is no reason to suppose that the survivors are older. This station is said to have been known as early as 1846, at which date the ground where they stand was grassy and fertile. These trees, if standing at that time, must assuredly have been in their infancy. The encroachment of the sea and subsequent change of conditions account well enough for the present decrepitude, but their general similarity in size and apparent age point rather to introduction than native growth.

South to Florida, Alabama, and Louisiana; west to Iowa, Kansas, and Texas.

Habit.—One of the Rhode Island trees measured 3 feet 11 inches girth at the base, and gradually tapered to a height of more than 40 feet (L. W. Russell). The trees at New Haven are 15-20 feet in height, with a trunk diameter of 6-10 inches, trunk and limbs much twisted by the winds. Their branches, beginning to put out at a height of 6-8 feet, lie in almost horizontal planes, forming a roundish, open head.

Bark.—Trunk in old trees dark, rough, deeply furrowed, separating into small, firm sections; large limbs dark reddish-brown; season's shoots green, turning to brown.

Winter Buds and Leaves.—Buds oblong, conical, short. Leaves simple, alternate, 3-6 inches long, about half as wide, dark green and mostly glossy above, somewhat lighter and minutely downy (at least when young) beneath, ovate to oval, entire; apex acute to acuminate; base acute, rounded or truncate; leafstalk short; stipules none.

Inflorescence.—June. Sterile and fertile flowers on separate or on the same trees; not conspicuous, axillary; sterile often in clusters, fertile solitary; calyx 4-6-parted; corolla 4-6-parted; about ½ inch long, pale yellow, thickish, urn-shaped, constricted [Pg 162] at the mouth and somewhat smaller in the sterile flowers; stamens 16 in the sterile flowers, in fertile flowers 8 or less, imperfect; styles 4, ovary 8-celled.

Fruit.—A berry, ripe in late fall, roundish, about an inch in diameter, larger farther south, with thick, spreading, persistent calyx, yellow to yellowish-brown, very astringent when immature, edible and agreeable to the taste after exposure to the frost; several-seeded.

Horticultural Value.—Hardy along the south shore of New England; prefers well-drained soil in open situations; free from disfiguring enemies; occasionally cultivated in nurseries but difficult to transplant. Propagated from seed.

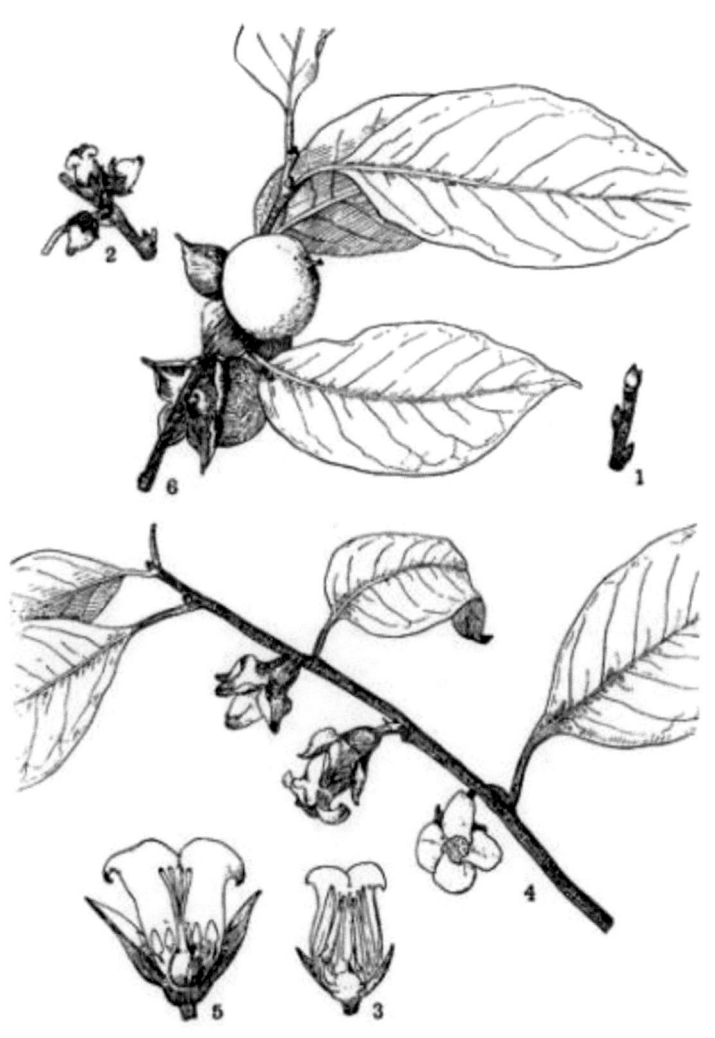

Plate LXXXII. — Diospyros Virginiana.

1. Winter buds.
2. Branch with sterile flowers.
3. Vertical section of sterile flower.
4. Branch with fertile flowers.
5. Section of fertile flower.
6. Fruiting branch.

OLEACEÆ. OLIVE FAMILY.

Fraxinus Americana, L.

White Ash.

Habitat and Range. — Rich or moist woods, fields and pastures, near streams.

Newfoundland and Nova Scotia to Ontario.

Maine, — very common, often forming large forest areas; in the other New England states, widely distributed, but seldom occurring in large masses.

South to Florida; west to Minnesota, Nebraska, Kansas, and Texas.

Habit. — A tall forest tree, 50-75 feet high, with a trunk diameter of 2-3 feet; rising in the rich bottom lands of the Ohio river 100 feet or more, often in the forest half its height [Pg 163] without a limb. In open ground the trunk, separating at a height of a few feet, throws off two or three large limbs, and is soon lost amid the slender, often gently curving branches, forming a rather open, rounded head widest at or near the base, with light and graceful foliage, and a stout, rather sparse, glabrous, and sometimes flattish spray.

Bark. — Bark of trunk in mature trees easily distinguishable at some distance by the characteristic gray color and uniform striation; ridges prominent, narrow, flattish, firm, without surface scales but with fine transverse seams; furrows fine and strong, sinuous, parallel or connecting at intervals; large limbs more or less furrowed; smaller branches smooth and grayish-green; season's shoots polished olive green; leaf-scars prominent.

Winter Buds and Leaves.—Buds short, rather prominent, smooth, dark or pale rusty brown. Leaves pinnately compound, opposite, 6-12 inches long; petiole smooth and grooved; leaflets 5-9, 2-5 inches long, deep green and smooth above, paler and smooth, or slightly pubescent (at least when young) beneath; ovate to lance-oblong, entire or somewhat toothed; apex pointed; base obtuse, rounded or sometimes acute; leaflet stalks short, smooth; stipules and stipels none.

Inflorescence.—May. In loose panicles from lateral or terminal buds of the previous season's shoots, sterile and fertile flowers for the most part on separate trees, numerous, inconspicuous; calyx in sterile flowers 4-toothed, petals none, stamens 2-4, anthers oblong; calyx in fertile flowers unequally 4-toothed or nearly entire, persistent; petals none, stamens none, pistil 1, style 1, stigma 2-cleft.

Fruit.—Ripening in early fall, and hanging in clusters into the winter; a samara or key 1-2 inches long, body nearly terete, marginless below, dilating from near the tip into a wing two or three times as long as the body.

Horticultural Value.—Hardy throughout New England; prefers a rich, moist, loamy soil, but grows in any well-drained situation; easily transplanted, usually obtainable in nurseries, and can be collected successfully. It is one of the most desirable native trees for landscape and street plantations, on [Pg 164] account of its rapid and clean growth, freedom from disease, moderate shade, and richly colored autumn foliage. As the leaves appear late in spring and fall early in autumn, it is desirable to plant with other trees of different habit. Propagated from seed.

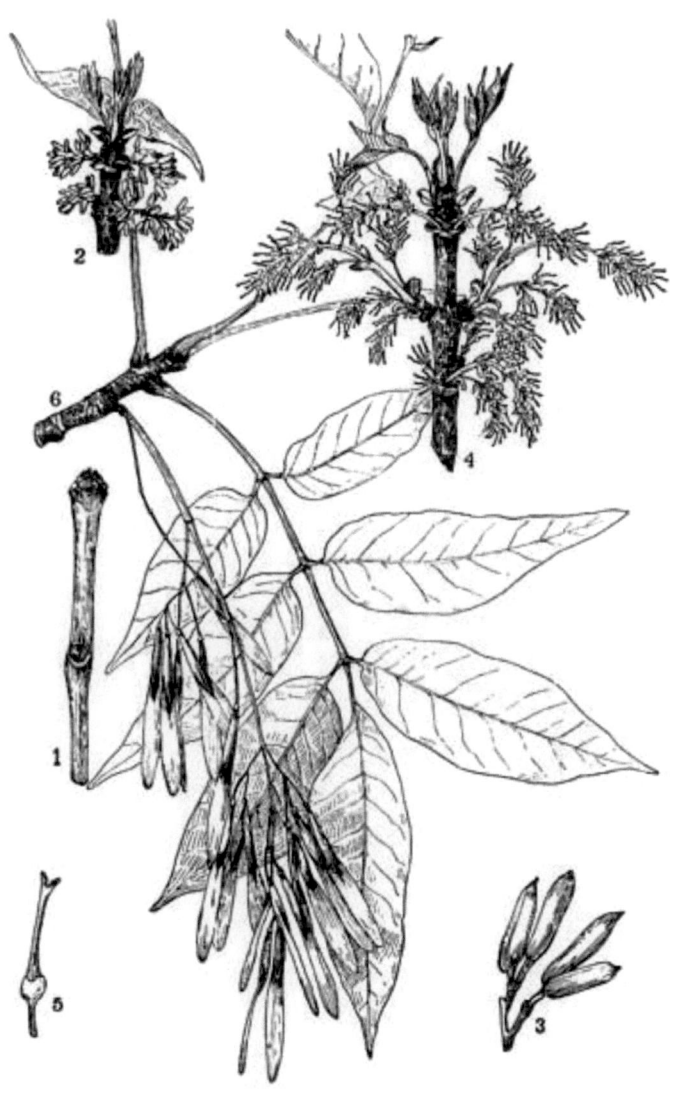

Plate LXXXIII. — Fraxinus Americana.

1. Winter buds.
2. Branch with sterile flowers.
3. Sterile flowers.
4. Branch with fertile flowers.
5. Fertile flower.
6. Fruiting branch.

Fraxinus Pennsylvanica, Marsh.

Fraxinus pubescens, Lam.

Red Ash. Brown Ash. River Ash.

Habitat and Range.—River banks, swampy lowlands, margins of streams and ponds.

New Brunswick to Manitoba.

Maine,—infrequent; New Hampshire,—occasional, extending as far north as Boscawen in the Merrimac valley; Vermont,—common along Lake Champlain and its tributaries (*Flora of Vermont*, 1900); occasional in other sections; Massachusetts and Rhode Island,—sparingly scattered throughout; Connecticut,—reported from East Hartford, Westville, Canaan, and Lisbon (J. N. Bishop).

South to Florida and Alabama; west to Dakota, Nebraska, Kansas, and Missouri.

Habit.—Medium-sized to large tree, 30-70 feet high, with trunk 1-3 feet in diameter; erect, branches spreading, broad-headed; in general appearance resembling the white ash.

Bark.—Trunk dark gray or brown, smooth in young trees, furrowed in old, furrows rather shallower than in the white ash; branches grayish; young shoots greenish-gray with a [Pg 165] rusty-velvety or scurfy pubescence lasting often into the second year.

Winter Buds and Leaves.—Buds rounded, dark reddish-brown, more or less downy, smaller than those of the white ash, partially covered by the swollen petiole. Leaves pinnately compound, opposite, 9-15 inches long; petiole short, downy, enlarged at base; leaflets 7-9, opposite, 3-5 inches long, about one half as wide, light green and smooth above, paler and more or less downy beneath; outline extremely variable, ovate, narrow-oblong, elliptical or sometimes obovate, entire or slightly toothed; apex acute to acuminate; base acute or rounded; leaflet stalks short, grooved, downy; stipules and stipels none.

Inflorescence.—May. Similar to that of the white ash.

Fruit.—Ripening in early fall, and hanging in clusters into the winter; samara or key about 1½ inches long; body of the fruit narrowly cylindrical, the edges gradually widening from about the center into linear or spatulate wings, obtuse or rounded at the ends, sometimes mucronate.

Horticultural Value. — Hardy throughout New England; grows readily in any good soil, but prefers a wet or moist, rich loam; almost as rapid growing when young as the white ash, and is not seriously affected by insects or fungous diseases; worthy of a place in landscape plantations and on streets, but not often found in nurseries; propagated from seed.

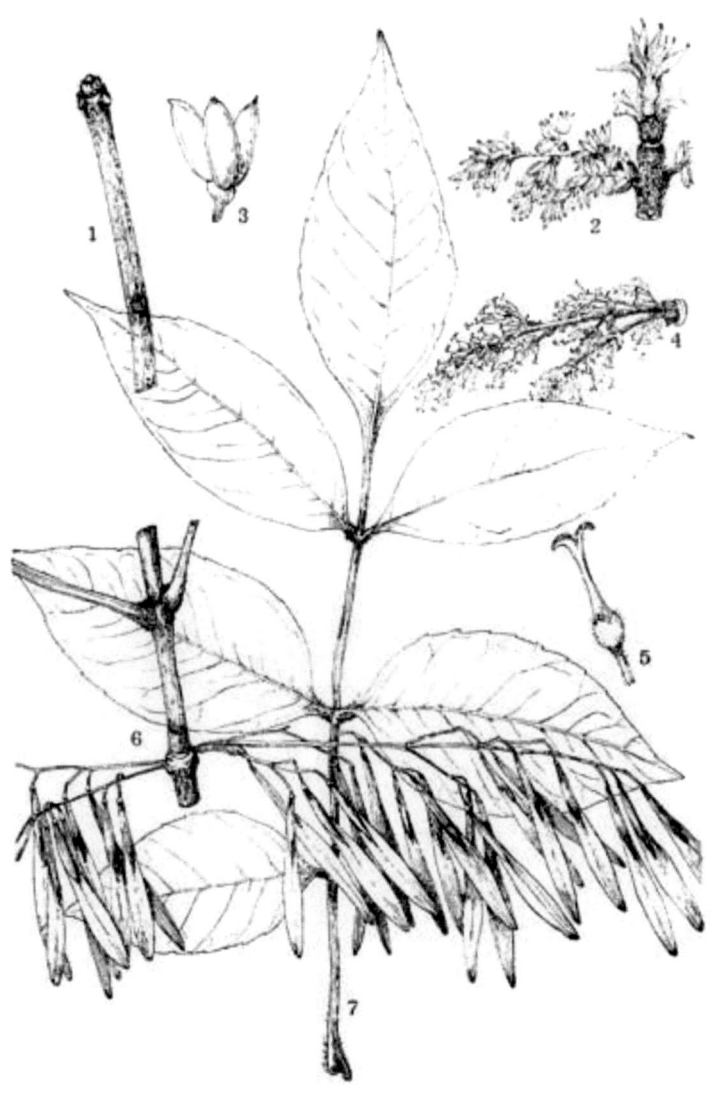

1

3

2

4

5

6

7

Plate LXXXIV. — Fraxinus Pennsylvanica.

1. Winter buds.
2. Branch with sterile flowers.
3. Sterile flowers.
4. Branch with fertile flowers.
5. Fertile flower.
6. Fruiting branch.
7. Mature leaf.

   [Pg 166]

Fraxinus Pennsylvanica, var. lanceolata, Sarg.

Fraxinus viridis, Michx. f. Fraxinus lanceolata, Borkh.

Green Ash.

River valleys and wet woods.

Ontario to Saskatchewan.

Maine,—common along the Penobscot river from Oldtown to Bangor; Vermont,—along Lake Champlain; Gardner's island, and the north end of South Hero; Rhode Island (Bailey); Connecticut,— frequent (J. N. Bishop, *Report of Connecticut Board of Agriculture*, 1895).

South along the mountains to Florida; west to the Rocky mountains.

The claims to specific distinction rest mainly upon the usual absence of pubescence from the young shoots, leaves and petioles, the color of the leaves (which is bright green above and scarcely less so beneath), the usually more distinct serratures above the center, and a rather more acuminate apex.

Apparently an extreme form of *F. pubescens*, connected with it by numerous intermediate forms through the entire range of the species.

436

Plate LXXXV.—Fraxinus Pennsylvanica, var. ceolata.

1. Winter buds.
2. Fruiting branch.

[Pg 167]

Fraxinus nigra, Marsh.

Fraxinus sambucifolia, Lam.

Black Ash. Swamp Ash. Basket Ash. Hoop Ash. Brown Ash.

Habitat and Range. — Wet woods, river bottoms, and swamps.

Anticosti through Ontario.

Maine, — common; New Hampshire, — south of the White mountains; Vermont, — common; Massachusetts, — more common in central and western sections; Rhode Island, — infrequent; Connecticut, — occasional throughout.

South to Delaware and Virginia; west to Arkansas and Missouri.

Habit. — A tall tree reaching a height of 60-80 feet, with a trunk diameter of 1-2 feet; attaining greater dimensions southward. In swamps, when shut in by other trees, the trunk is straight, very slender, scarcely tapering to point of branching, in open situations under favorable conditions forming a large, round, open head. Easily distinguished from the other ashes by its sessile leaflets.

Bark. — Bark of trunk a soft ash-gray, in old trees marked by parallel ridges separating into fine, thin, close flakes; limbs light gray, rough-warted, the smaller with conspicuous leaf-scars; season's shoots olive green, stout; flattened at apex, with small, black, vertical dots.

Winter Buds and Leaves. — Buds roundish, pointed, very dark, the terminal ⅛ inch long. Leaves compound, opposite, 12-15 inches long; stipules none; stem grooved and smooth; leaflets 7-11, more frequently 9, 3-5 inches long, 1½-2 inches wide, green on both sides, lighter beneath and more or less hairy on the veins; outline variable, more usually oblong-lanceolate, sharply serrate; apex acuminate; base obtuse to rounded, sessile except the odd leaflets; stipels none.

Inflorescence. — May. Appearing before the leaves in loose panicles from lateral or terminal buds of the preceding season, sterile and fertile flowers on different trees; bracted; calyx none; petals none. [Pg 168]

Fruit. — August to September. Samaras, in panicles, rather more than 1 inch long, rounded at both ends: body entirely surrounded by the wing.

Horticultural Value. — Hardy throughout New England; grows in any good soil, but prefers swamp or wet land. Its very tall, slender habit makes it a useful tree in some positions, but it is not readily obtainable in nurseries and is seldom used. Propagated from the seed.

Plate LXXXVI. — Fraxinus nigra.

1. Winter buds.
2. Branch with sterile flowers.
3. Sterile flower.
4. Branch with fertile flowers.
5. Fertile flower.
6. Fruiting branch.
7. Fruit.

CAPRIFOLIACEÆ. HONEYSUCKLE FAMILY.

Viburnum Lentago, L.

Sheep Berry. Sweet Viburnum. Nanny Plum.

Habitat and Range. — Rich woods, thickets, river valleys, along fences.

Province of Quebec to Saskatchewan.

Frequent throughout New England.

South along the mountains to Georgia and Kentucky; west to Minnesota, Nebraska, and Missouri.

Habit. — A shrub or small tree, 10-25 feet in height with numerous branches forming a wide-spreading, compact rounded head; conspicuous by rich foliage, profuse, fragrant yellowish-white flowers, and long, drooping clusters of crimson fruit which deepen to a rich purple when fully ripe.

Bark. — Trunk and larger branches dark purplish or reddish brown, separating in old trees into small, firm sections; branchlets grayish-brown; season's shoots reddish-brown, dotted, more or less scurfy. [Pg 169]

Winter Buds and Leaves. — Leaf-buds long, narrow, covered with scurfy, brown, leaf-like scales; flower-buds much longer, swollen at the base, with two leaf-like scales extended into a long, spire-like point. Leaves simple, opposite, 2-4 inches long, upper surface bright green, lower paler and set with rusty scales, ovate to oblong-ovate or orbicular, sharply and finely serrate, smooth, tapered or abruptly pointed; base acute to rounded or truncate; stem slender, wavy-margined, channeled above; stipules none.

Inflorescence. — May or early June. Terminal, in broad, flat-topped, compound, sessile cymes; calyx-tube adherent to the ovary, 5-toothed; corolla white, salver-shaped, segments 5, oval, reflexed; stamens 5, projecting, anthers yellow; pistil truncate.

Fruit. — Profuse, in clusters; drupes ½ inch long, oval, crimson when ripening, deep purple when fully ripe, edible, sweet: stone flat, oval, rough, obscurely striate lengthwise.

Horticultural Value. — Hardy throughout New England; prefers a rich soil in open places or in light shade. Its showy flowers, healthy foliage, and vigorous growth make it a desirable plant for high shrub plantations, and as an undergrowth in open woods. Offered for sale by collectors and occasionally by nurserymen; easily transplanted; propagated from seed or from cuttings.

Plate LXXXVII. — Viburnum Lentago.

1. Winter buds.
2. Flowering branch.
3. Flower.
4. Flower, side view.
5. Flower with petals and stamens removed.
6. Fruiting branch.

[Pg 171] [Pg 170]

APPENDIX.

The range of several trees as given in the text has been extended by discoveries made during the summer of 1901, but reported too late for incorporation in its proper place.

Populus balsamifera, L., var. *candicans*, Gray. — One of the commonest and stateliest trees in the alluvium of the Connecticut and the Cold rivers; with negundo, river maple, and white and slippery elm, forming a tall and dense forest along the Connecticut at the foot of Fall mountain, and opposite Bellows Falls. The densely pubescent petioles and the ciliate margins of the broad cordate leaves at once distinguish this tree from the usually smaller but more common *P. balsamifera* ("Some Trees and Shrubs of Western Cheshire County, N. H." Mr. M. L. Fernald, in *Rhodora*, III, 233).

The above is the *Populus candicans*, Ait., of the text.

Salix discolor, Muhl. — There are many fine trees at Fort Kent, Maine, one with trunk 13 inches in diameter. (M. L. Fernald *in lit.*, September, 1901.)

Salix balsamifera, Barrett. — A handsome tree at Fort Kent, 25-30 feet high, with trunk 4-6 inches in diameter. (M. L. Fernald *in lit.*, September, 1901.)

Cratægus Crus-Galli, L. — Nantucket, Massachusetts. Young trees were set out in 1830, enclosing an oblong of about an acre and a half. The most flourishing of these have obtained a height of about 30 feet and a trunk diameter near the ground of 10-12 inches. Now established, probably through the agency of birds, along swamps and upon hill-slopes. (L. L. D.)

Prunus Americana, Marsh.—One clump of small trees in a thicket at Alstead Centre, N. H., has the characteristic spherical fruit of this species. *P. nigra*, Ait., with oblong, laterally flattened fruit, is abundant. (*Rhodora*, III, 234.) [Pg 172]

Acer Saccharum, Marsh., var. *barbatum*, Trelease.—Characteristic trees (Cheshire County, N. H.), with small, firm, deep green, three-lobed leaves, appear very distinct, but many transitions are noted between this and the typical *Acer Saccharum*. (*Rhodora*, III, 234.)

Acer Saccharum, Marsh., var. *nigrum*, Britton.—Occasional in alluvium of the Cold river (Cheshire county, N. H.). The large, dark green, "flabby" leaves, with closed sinuses and with densely pubescent petioles and lower surfaces, quickly distinguish this tree from the ordinary forms of the sugar maple. (*Rhodora*, III. 234.)

Fraxinus Pennsylvanica. Marsh., var. *lanceolata*, Sarg.—Common along the Connecticut at Walpole, N. H. (M. L. Fernald *in lit.*, September, 1901.) [Pg 173]

GLOSSARY.

Abortive. Defective or barren, through non-development of a part.

Acuminate. Long-pointed.

Acute. Ending with a sharp but not prolonged point.

Adherent. Growing fast to; adnate anther, attached for its whole length to the ovary.

Adnate. Essentially same as adherent, with the added idea of congenital adhesion.

Aggregate fruits. Formed by crowding together all the carpels of the same flower; as in the blackberry.

Ament. Name given to such flower-clusters as those of the willow, birch, poplar, etc.

Anther. The part of the stamen which bears the pollen.

Appressed. Lying close against another organ.

Ascending. Rising upward, or obliquely upward.

Axil. Angle formed on the upper side between the leaf stem or flower stem and the branch from which it springs.

Bract. Reduced leaf subtending a flower or flower-cluster.

Branches, primary. The leading or main branches thrown out directly from the trunk, giving a general shape to the head.

Branches, secondary. Never directly from the trunk but from other branches.

Buttressed. Supported against strain in any direction by a conspicuous ridge-like enlargement of the trunk vertically to the roots. Several of these buttresses often give a tree a square appearance.

Caducous. Dropping off very early after development.

Calyx. The outer set of the leaves of the flower.

Campanulate. Bell-shaped.

Capitate. Head-shaped or collected in a head.

Capsule. A dry compound fruit.

Carpel. A simple pistil.

Catkin. See ament.

Ciliate. Margin with hairs or bristles.

Coherent. One organ uniting with another.

Compound. See leaf, ovary, etc. [Pg 174]

Connate. Similar organs, more or less grown together.

Connective. The part of the anther connecting its two cells.

Coriaceous. Thick, leathery in texture.

Corolla. Leaves of the flower within the calyx.

Corymb. That sort of flower-cluster in which the flower stems arranged along the central axis elongate, forming a broad convex or level top, the flowers opening successively from the outer edge towards the center.

Crenate. Edge with rounded teeth.

Crenulate. Edge with small rounded teeth.

Cyme. Flat-topped or convex flower-cluster, the central flower opening first; blossoming outward.

Deciduous. Falling off, as leaves in autumn, or calyx and corolla before fruit grows.

Declining. Bent downwards.

Decurrent. Leaves prolonged on the stem beneath the insertion: branchlets springing out beneath the point of furcation, as the feathering along the trunk of elms, etc.

Dentate. With teeth pointing outwards.

Disk. Central part of a head of flowers; fleshy expansion of the receptacle of a flower; any rounded, flat surface.

Drupe. A stone fruit; soft externally with a stone at the center, as the cherry and peach.

Erose. Eroded, as if gnawed.

Exserted. Protruding, projecting out of.

Falcate. Scythe-shaped.

Fertile. Flowers containing the pistil, capable of producing fruit. Anthers in such blossoms, if any, are generally abortive.

Fibrovascular. Bundle or tissue, formed of wood fibers, ducts, etc.

Filament. Part of stamen supporting anther.

Fungus. A division of cryptogamous plants, including mushrooms, etc.

Furcation. Branching.

Glabrous. Smooth without hairiness or roughness.

Glandular. Bearing glands or appendages having the appearance of glands.

Glaucous. Covered with a bloom: bluish hoary.

Globose or **globous.** Spherical or nearly so.

Habit. The general appearance of a plant.

Habitat. The place where a plant naturally grows, as in swamps, in water, upon dry hillsides, etc. [Pg 175]

Hybrid. A cross between two species.

Imbricated. Overlapping.

Inflorescence. Mode of disposition of flowers; sometimes applied to the flower-cluster itself.

Involucre. Bracts subtending a flower or a cluster of flowers.

Keeled. Having a central dorsal ridge like the keel of a boat.

Key. A winged fruit; a samara.

Lacerate. Irregularly cleft, as if torn.

Lanceolate. Lance-shaped, broadest above the base, gradually narrowing to the apex.

Leaf. Consisting when botanically complete of a blade, usually flat, a footstalk and two appendages at base of the footstalk; often consisting of blade only.

Leaf, compound. Having two to many distinct blades on a common leafstalk or rachis. These blades may be sessile or have leafstalks of their own.

Leaf, pinnately compound. With the leaflets arranged along the sides of the rachis.

Leaf, palmately compound. With leaflets all standing on summit of petiole.

Leaf-cushions. Organs resembling persistent decurrent footstalks, upon which leaves of spruces, etc., stand; sterigmata.

Leaf-scar. The scar left on the twig where the petiole was attached.

Lenticel. Externally appearing upon the bark as spots, warts, and perpendicular or transverse lines.

Linear. Long and narrow with sides nearly parallel.

Monopetalous. Having petals more or less united.

Mucronate. Abruptly tipped with a small, sharp point.

Nerved. Having prominent unbranched ribs or veins.

Obcordate. Inversely heart shaped.

Obovate. Ovate with the broader end towards the apex.

Obtuse. Blunt or rounded at the end.

Orbicular. Having a circular or nearly circular outline.

Ovary. The part of the pistil containing the ovules.

Ovoid. A solid with an oval or ovate outline.

Ovuliferous. Bearing ovules.

Panicle. General term for any loose and irregular flower-cluster, commonly of the racemose type, with pedicellate flowers.

Pedicel. The stalk of a single flower in the ultimate divisions of an inflorescence.

Peduncle. The stem of a solitary flower or of a cluster. [Pg 176]

Perfect. Having both pistils and stamens.

Perianth. The floral envelope consisting of calyx, corolla, or both.

Persistent. Not falling for a long time.

Petal. A division of the corolla.

Petiole. The stalk of a leaf.

Petiolule. The stalk of a leaflet in a compound leaf.

Pistil. The seed-bearing organ of the flower.

Pistillate. Provided with pistils; usually applied to flowers without stamens.

Pollen. The fertilizing grains contained in the anthers.

Puberulent. Minutely pubescent.

Pubescent. Covered with short soft or downy hairs.

Raceme. A simple cluster of pediceled flowers upon a common axis.

Rachis. The main axis of a compound leaf, of a raceme or of a spike.

Ramification. Branching.

Range. The geographical extent and limits of a species.

Reflexed. Turned backward.

Reticulated. Netted; in the form of a network.

Revolute. Rolled backward from the margin or apex.

Samara. Key fruit; winged fruit, like that of the ash or maple.

Scarf-bark. The thin, outermost layer which often peels off.

Segment. One of the divisions into which a plane organ, such as a leaf, may be divided.

Sepal. A calyx leaf.

Serrate. With teeth inclining forward.

Serrulate. With small teeth inclining forward.

Sessile. Not stalked, as when the leaf blade or flower rests directly upon the twig.

Simple leaf. Not compound, having one blade not jointed with its stem.

Sinuate. Strongly wavy-margined.

Sinus. Interval between two lobes or divisions of a leaf; sometimes sharp-angular, sometimes rounded.

Spatulate. Gradually narrowed downward from a rounded summit.

Spike. A cluster of sessile or nearly sessile lateral flowers on an elongated axis.

Spray. The smaller branches and ultimate branchlets of a tree taken as a whole.

Stamens. The pollen-bearing organs of a flower, each stamen consisting of a filament (stem) and anther which contains the pollen.

Staminate. Having stamens. [Pg 177]

Sterile. Variously applied: to flowers with stamens only; to stamens without anthers; to anthers without pollen; to ovaries not producing seed, etc.

Stigma. Part of pistil which receives the pollen.

Stipels. Appendages to a leaflet, analogous to the stipules of a leaf.

Stipules. Appendages of a leaf, usually at the point of insertion.

Striate. Streaked, or very finely ridged lengthwise.

Style. Part of pistil uniting ovary with stigma; often wanting.

Sucker. A shoot of subterranean origin.

Suture. The line of union between parts which have grown together; most often used with reference to the line along which an ovary opens.

Terete. Cylindrical.

Ternate. In threes.

Tomentose. Densely pubescent or woolly.

Truncate. As if cut off at the end.

Umbel. An inflorescence in which the flower stems spring from the same point like the rays of an umbrella.

Verticillate. Arranged in a circle round an axis; whorled.

Villose or **villous.** With long, soft hairs.

Whorl. Arranged in a circle about an axis. [Pg 179] [Pg 178]